THE PERFECT SCENT

THE PERFECT SCENT

A Year Inside

the Perfume Industry

in Paris and New York

Chandler Burr

Henry Holt and Company
New York

Henry Holt and Company, LLC
Publishers since 1866
175 Fifth Avenue
New York, New York 10010
www.henryholt.com

Henry Holt® *and* 🛡® *are registered trademarks of Henry Holt and Company, LLC.*

Copyright © 2007 by Chandler Burr
All rights reserved.

Distributed in Canada by H. B. Fenn and Company Ltd.

Library of Congress Cataloging-in-Publication Data

Burr, Chandler, date.
   The perfect scent : a year inside the perfume industry in Paris and New York / Chandler Burr.—1st ed.
      p. cm.
   ISBN-13: 978-0-8050-8037-7
   ISBN-10: 0-8050-8037-6
      1. Perfumes industry—Popular works.   2. Perfumes—Popular works.   I. Title.
HD9999.P3932B87   2008
338.7′66854—dc22   2007024460

Henry Holt books are available for special promotions and premiums.
For details contact: Director, Special Markets.

First Edition 2008

Designed by Meryl Sussman Levavi

Printed in the United States of America

10   9   8   7   6   5   4   3   2   1

This book is dedicated to
Joseph Andrew Tomkiewicz from Wisconsin.
The best friend a guy could reasonably ask for.

Ce qui vient au monde pour ne rien troubler,
ne mérite ni égard ni patience.

WHAT COMES INTO THE WORLD TO DISTURB NOTHING
MERITS NEITHER ATTENTION NOR PATIENCE.

—RENÉ CHAR (1948)

ART DOES NOT REPRODUCE THE VISIBLE;
RATHER, IT MAKES VISIBLE.

—PAUL KLEE

Les absents ont toujours tort.

THOSE WHO ARE ABSENT ARE ALWAYS WRONG.

—PHILIPPE NÉRICAULT DESTOUCHES
(1717)

O N June 9, 2004, just before 5:00 P.M., Jean-Claude Ellena was being driven to a meeting at the offices of
Parfums Hermès in Pantin, just outside the *périphérique*
to the northeast of Paris. Ellena was a famous ghost, a member of
an elite group of perfumers who create fragrances sold under the
names of designers and luxury houses while keeping assiduously to
the shadows. But he was just at the point of becoming particularly,
and rather extraordinarily, visible to the world. He was on his way
to Hermès to submit his first *essais*, his olfactory sketches, for an
important scent he was creating.

Paris was enjoying a spell of Los Angeles–like weather. You
could look from the top of rue Ménilmontant down over the Cen-
tre Georges Pompidou's industrial modernism all the way to the
Eutelsat balloon floating over the Parc André Citroën. In the deep-
cobalt summer sky, the cloud of aerosolized filth from the Paris

traffic hovered in the blue air. The sun shone brightly. The Parisians walked around wearing black, smoking cigarettes, exhaling ashen fumes into the air, and throwing the butts and packets onto streets where Africans in cotton *bleus de travail* uniforms swept them into sewers.

From his car, Ellena looked out at the bus stops. It seemed as if every single one featured an ad for Chanel's latest feminine perfume, *Chance*. It was a bit startling. The car crossed an avenue, stopped at a light: *Chance*. It turned right: *Chance*. Ellena looked left; from every vantage the publicity image of a wispy blond girl floated spectrally over the round metallic glass *Chance* bottle. This display represented a breathtaking marketing outlay. If you were in the perfume industry, if you were the competition—say, another immaculate luxury house like Hermès—you might not show any reaction. You might smile, eyes focused just beyond the ads. But you would register them as they slid by your car, this show of Chanel's stunning power, a silken reminder of the might of this billion-dollar titanium luxury machine. The bus ads were not a campaign. They were a statement. "We are here." Their ubiquitousness was profoundly intimidating. This was the intention.

Hermès had, in fact, two responses. The first was the three small vials in Ellena's pocket, each containing a pale golden-colored scent. The second was Ellena himself.

⁓◎

Across the Atlantic not many months later, at 1:00 P.M. on October 29, 2004, the actress Sarah Jessica Parker arrived at the offices of her agent, Peter Hess, at Creative Artists Agency at 162 Fifth Avenue in New York City. She was there to meet representatives from Coty, Inc., the international perfume licensing corporation, whose headquarters were just up the street. Parker and her representatives would be discussing the final details of a contract for the creation of a perfume that would bear her name.

They met in one of the white CAA conference rooms. Along with Hess, Parker's rep Ina Treciokas from the public relations agency IDPR was present. The Coty contingent numbered four, all perfume industry executives and "creatives" (as those in charge of developing a perfume are called in the industry). There was excellent sushi and a big bowl of popcorn, a neat line of drinks, and bowls of ice. Parker was dressed in relaxed style—jeans and a T-shirt—but she was quite alert to the significance of the meeting and to the variables at play.

Parker had for years been a star on stage and on screen, but she was as aware as anyone of the risks of attempting to transfer the mercurial, amorphous good of celebrity to other domains. In both a symbolic and a literal sense, she was funding this project with her public equity. But she had for years wanted to create a perfume— "dreamed of it," as she expressed it eagerly to the Coty team that day. Peter Hess and CAA had been pursuing it for her, making the contacts, talking to the players in the perfume world—the luxury juggernauts like the Lauders and LVMHs, with their brands and labs and marketing armies—and Hess had found the process far from easy; the perfume industry is brutal, and the financial stakes increasingly high. Yet Coty was interested in Parker, and the lawyers— Coty's and the star's—had been working on the contract for many months. It had been a complicated negotiation.

Hess naturally shared Parker's concerns. Were she to give Coty the license to her name and her public identity, the project would entail years of effort on her part and that of the Coty team that would develop the scent with her, millions of dollars put into the launch and a massive promotional campaign, and the risk of her image and reputation.

It would also require of Parker a special, and rather unusual, form of participation. During the development of the scent, she would assume the position known in the industry as artistic director. She would have to guide the perfumers who would build her scent.

She would be responsible for directing them toward a precise olfactory representation of an idea of a perfume she already had in her head. Parker had never played the role before—it was the perfumers who understood mixing rose absolute with dihydrojasmonate, not she—and she didn't, truth be told, know exactly what to expect.

⁓◉

Between 2004 and 2006, I reported these two stories, one for *The New Yorker*, the other for *The New York Times*. Both were intimate behind-the-scenes accounts of two very different people creating two very different perfumes. Ellena's scent was built at and for Hermès, among the last family-owned exclusive luxury goods houses in France, based in an eighteenth-century shop on the rue du Faubourg. It was created in Paris and in Grasse, France's traditional capital of scent. Parker's fragrance was made under the corporate aegis of one of the largest commercial producers of perfume in the world, a company headquartered in a skyscraper in New York City.

The first perfume was *Un Jardin sur le Nil*. The other was *Sarah Jessica Parker Lovely*.

It happened that I fell into both of these stories—they found me, each one in its own particular way. Both of these scents were built to be launched on the $31-billion international perfume market, and in the course of reporting on their respective creation processes, I spent two years inside this industry, one of the most insular, glamorous, strange, paranoid, idiosyncratic, irrational, and lucrative of worlds.

I am the perfume critic for *The New York Times*, but I am not a visceral perfume obsessive. Some people want me to be, but I'm not. Fundamentally I'm a reporter and critic whose job is to write on perfume—the business, industry, and personalities, and of course the works of commercial art they produce, the perfumes. It's a professional beat. At the same time, writing about perfume has held a real, and I will admit visceral, surprise, which is that I am now conscious of experiencing the world more deeply and vividly than I'd ever thought possible. Many people situate themselves by sight; they marvel at scenic vistas, take photos, draw pictures, recall images. In this job I find my brain recording time and place in scent. I remember places by smell.

In travel, smell is our best, most reliable landmark. Researchers have found that our ability to recall a specific scent surpasses even our ability to recall what we've seen. Show photos to people, then show them the photos months later; it's estimated that visual recall is about 50 percent after four months. Trygg Engen, a professor of psychology at Brown University, found that people recall smells with 65 percent accuracy after a year. If you've been in Africa or Asia or Latin America for any significant period of time and then return home, open your suitcase and take out the clothes, and the aroma places you. You're in Nairobi or the bush, Bangkok's center, the beach in Rio. Photos can't do that. Smell transports us, beautifully, strongly, insistently. The smell of my childhood was South Texas, but also the aroma of travel, of jet fuel, the synthetic carpets of airports, and the recycled air of planes. My grandmother, Marjorie Langston Stewart, lived in the Corpus Christi of the 1960s, two blocks from the Gulf of Mexico. While I remember her voice and face, she actually exists in my memory as a fragrance: of fresh citrus and the green leaves of the poisonous Texas oleander she warned my sister and me never to touch, the hot wet breezes of the Gulf of Mexico off fishing boats, the clean, rich Victorian smells of the England she grew up in that scented her house and the scent of her immense, powerful, white 1958 Pontiac's interior. The years I spent following these two stories were mapped in scent in just this way, and I recall its chapters by their smells.

The main actors in each story were utterly different—a French perfumer and an American movie star—and the companies and contexts in which they worked were dissimilar in a thousand ways, but both stories began with a problem.

﹋◉

At Hermès, the problem was simple to identify, tough to solve.

Hermès is as close to an immaculate brand as it is possible to get. In the business world, the name commands absolute respect. But while consumers from Paris to Osaka faithfully snapped up astoundingly expensive Hermès silks, clothes, and leather bags, the Hermès family was quietly aware of a weakness in the house: its perfumes.

Clients on Fifth Avenue and the Ginza, avenue Montaigne and Via Monte Napoleone were reaching for bottles of Chanel, Armani, Calvin Klein, and Dior well before they reached for an Hermès scent. For a luxury house, a perfume problem is not just an image issue. It goes inevitably to the heart of your business. Because of its profit margins and its massive distribution, perfume is a crucial money generator for almost all the high-end houses. It's an open secret that fragrance is essential to the financial health of most of the world's luxury brands. A man I know once sat next to Yves Saint Laurent at a Paris dinner party. He asked, "What portion of Yves Saint Laurent revenues are accounted for by perfume?" Saint Laurent replied, "Eighty-three point five."

By 2004, Hermès (the name's French pronunciation is closest to a combination of "air-*mess*" and "air-*mezz*") was ready to confront the perfume problem. Jean-Louis Dumas-Hermès was the head of the family, which owned 75 percent of the house. Dumas was elegant and refined and his personal worth was around $1.5 billion, and if he presented a certain gentleness that overlay a steely control, he was remarkably free of the arrogance that haute Parisian culture and great wealth could have produced in him. In 2004, Dumas hired a new president of Parfums Hermès, Véronique Gautier. Gautier had a significant reputation.

Dark-haired and dark-eyed, opinionated and direct, Gautier was a veteran of perfume operations at Chanel and Cartier. She was known for two skills: crafting grand strategies and imposing, with an iron will, the decisions needed to get them into place. Gautier had determinedly big vision, and she was relentlessly daring, both qualities Dumas needed. Gautier had, naturally enough, left in her professional wake a division of opinions. "Véronique is admired rather than liked," one of her French competitors, a woman, said with a hard look. Another gave a different assessment: "She knows the business as well as anyone, and she's competent and strong. Jean-Louis will need that."

The figures in front of Gautier were relatively straightforward.

In 2003, the previous year, the Italian jeweler Bulgari had had perfume sales totaling €136,700,000. This represented 18 percent of Bulgari's total revenues of €759,300,000. Hermès in the same year had sold €54 million in perfume—less than half of Bulgari's sales—on a total of €1.23 billion, which meant perfume was only 4.4 percent of Hermès's business.

Dumas and Gautier were exquisitely conscious of the perfume referred to by some in the industry as *le monstre*—the monster—Chanel's *Chanel No. 5*. Here was a ninety-year-old fragrance always at the top of the international bestseller lists, an institution whose 2003 sales had been an astonishing €180 million. Hermès had a collection that included excellent scents like *Calèche* and *24 Faubourg*, but their sales didn't even touch those of the fabled *No. 5*. Hermès had a waiting list for its $10,000 Kelly bags, yes, but these Hermès bags consistently carried more Chanel perfume. The question for Dumas and Gautier was: why?

Start with the open secret of the industry, which is that the perfumes purchased from Donna Karan, Ralph Lauren, or Giorgio Armani are not created by these designers. Domenico Dolce and Marc Jacobs don't make their own perfumes. They don't know how. They never have. In fact almost none of the people whose names go on the boxes have ever touched a scent raw material in their lives. All the fragrances in the world are made by an army of professional ghosts called perfumers, who are hired by the designers and brands. The ghosts live in a sort of netherworld carefully hidden from view by the designers' marketing machines. They work primarily for several international scent-maker corporations, the "Big Boys"— Givaudan and Firmenich (both Swiss), IFF (American), Symrise (German), and Takasago (Japanese)—plus the smaller players like Robertet, Drom, Fragrance Resources, Mane, and Belmay. Of them, the fashion houses do not speak publicly. In fact, most of them spend large amounts of money on public relations agencies in Milan, Paris, and New York in order explicitly to create the impression that the perfumes come from the designers.

Estée Lauder created none of her perfumes. A huge international corporation called IFF did. International Flavors & Fragrances is based in New York. Lauder gave the IFF perfumers concepts, she guided the scents to finalization, and she put her name on the scents they made, though not her real name, Josephine Esther Mentzer, nor the real names of the people who actually built the juices for her.

Youth-Dew, Estée Lauder's first perfume, in 1953, was made by the IFF perfumer Josephine Catapano. *Youth-Dew* started as a simple bath oil, just a gift Lauder gave to her clients. Lauder was unknown then, but IFF believed in her. Betty Busse made *Estée*, the legendary Francis Camail built *Aliage* (the following year he would create *Charlie* for Revlon), and the equally legendary Sophia Grojsman (*Trésor* for Lancôme and the beautiful *Jaipur* for Boucheron) did Lauder's *White Linen*, a landmark in everyone's view. *Private Collection* was the creation of Vince Marcello, *Beautiful* was made by five different IFF perfumers. Leonard Lauder gave credit for his multibillion-dollar success to Ernest Shiftan, IFF's chief perfumer who encouraged and built a generation of great American scent artists. Estée was, perfumers note, demanding and involved; she had taste and she had vision, and she closely creative-directed the scents they made. But perfumers also note (although they never do so on the record) that saying she created her own perfumes would be, as the perfume expert Michael Edwards phrased it, like saying Pope Julius painted the Sistine Chapel.

This arrangement is standard industry-wide practice. In 1947, Christian Dior's first, *Miss Dior*, was made by two Givaudan perfumers, Jean Carles and Paul Vacher. In 2007, the summer launch by Giorgio Armani, who is well-known for wanting the public to believe he makes his own perfumes, was built for him by the perfumers Francis Kurkdjian and Françoise Caron.

The arrangement is an uneasy one in all sorts of ways. How is an outside perfumer to incarnate a house in scent? What does the perfumer know of that house's aesthetic, taste, or style? Nothing, usually. Here was Hermès, founded as saddle and harness makers on the rue Basse-du-Rempart in 1837 (and hyperconscious of that distinguished date). Hermès was French, which meant that above all it was proud, obsessed with its craftsmanship and its pedigree, a house for whom coherence was a golden rule, and to create its perfumes it went to strangers? The quality of Hermès saddles was reflected in the leather in Hermès belts, all of it a seamless gleaming perfection. Except its fragrances. And this was the thing about *le monstre*.

Only one house did not employ the Big Boys and their perfumers: Chanel. Chanel had Polge. And Jacques Polge, Chanel's in-house perfumer, directed its perfume collection with precision. He was part of the house's genetics. Polge was only the third Chanel perfumer. (The first had been, from 1920, Ernest Beaux, who had created among others *No. 5* and *No. 22* under Coco Chanel's direct artistic direction. The second, Henri Robert, was author of *Cristalle* and *Pour Monsieur*.) Polge had authored *Coco* and *Allure* and, in 2001, *Coco Mademoiselle*, which joined the others on the bestseller lists. In him, Chanel had tradition, institutional memory, a coherent aesthetic. And although an in-house perfumer was an expensive proposition, Hermès needed all those things. Of course Dumas and Gautier wanted commercial success. But perhaps more important, they wanted a scent collection with the elegance and coherence of their leather bags and silk ties. They wanted Hermès perfumes, which was to say they wanted beautifully constructed fragrances carefully built by artisans' hands. They wanted perfumes whose exquisite purity distilled Hermès and were worthy of their smooth Gallic pride.

The answer to this, they decided, was a particular perfumer named Jean-Claude Ellena. Gautier flew down to Grasse, where Ellena lived, to make the proposal. She and Dumas already had plans for the initial Hermès perfume Ellena would make, but first they had to get him on board.

⁓⊚

My entry into the second story began when I got a call from Belinda Arnold, Coty's director of public relations at Coty's New York headquarters at Park Avenue and East Thirty-third Street.

Coty researches, develops, launches, and manages its brands. It claims to be the world's largest fragrance company, with annual net sales of $2.9 billion. (It doesn't publish the gross; Coty's principal competitors, LVMH, Estée Lauder, and L'Oréal, also lay claim to number one status, which depends a bit on accounting and a bit on exchange rates.) Coty operates in over twenty-five countries. Since Coty's playing field is global, its strategy calculates interests in markets across the planet. It owns Davidoff, which is minor in North America and a huge player in Europe, Asia, and the Middle East, and Coty continually acquires licenses to fortify its positions and expand into new markets. It also enters into strategic distribution partnerships. The perfume lines Nina Ricci, Carolina Herrera, Prada, Paco Rabanne, and Comme des Garçons are owned by a Coty competitor, the Spanish company Puig, but Coty distributes these brands in the United States.

Where Hermès makes all its own products itself, with its own designers, artisans, and raw materials—except its perfumes, and hiring Ellena was the effort to remedy this—Coty is a licensee, which is to say that it contracts with brands that are not part of itself and makes these brands' products using other people's materials. This is the way the large-scale perfume industry works. In 2005, when Belinda Arnold approached me, Coty, between its two divisions—"Prestige" and "Beauty"—was already huge and owned or licensed around thirty-five brands. It had celebrities—Gwen Stefani, Céline Dion, Shania Twain, David and Victoria Beckham, Kate Moss, Kylie Minogue, and the Olsen twins—and some brands that were huge in Europe like *Jil Sander, Joop, Lancaster,* and *Rimmel.* It had the supercommercial—from the license for Kimora Lee Simmons's Baby Phat line that produced the hideous *Goddess* to the license for the TV series *Desperate Housewives* that led to a pleasant commercial feminine of the same name—and it owned upscale antique brands like Pierre Cardin. It had the purest pop-culture licenses; Chupa Chups, the candy company, had lent its name to a perfume that Coty had had built. And Coty had just bought Unilever's entire fragrance division, which meant it had bought the licenses of several high-end designers from Calvin Klein and Marc Jacobs to Vera Wang, Kenneth Cole, and Vivienne Westwood, as well as Cerruti and Chloé.

In a difference partly theological, partly quite real, where Hermès was grappling with making its products authentic, Coty was grappling with making its products legitimate. In a sense this was due to one of the most astounding commercial successes the business world had seen in decades, a trend created almost single-handedly by Coty.

Celebrity perfumes caught everyone unawares except Catherine Walsh. Walsh arguably created them. Then senior vice president for cosmetics and American licenses at Coty Lancaster, Walsh, with an immense amount of work and risk, put together the deal that signed Jennifer Lopez. When Lopez's *Glow* hit the market in 2002, it exploded, selling 8 million bottles in a few short years. The modern incarnation of the celebrity perfume was born. The question for the celebrity perfume is, of course, how much of the celebrity is actually in the bottle. Did the marquee name have anything to do with its creation at all, and (moreover) how did the development process— a somewhat delicate dance between licensee and star—lead to a bottle of perfume? Belinda Arnold's job was a bit simpler; when the juices launched, she merely had to get the word out.

Belinda is one in an amazing army of young Manhattan PR women, all attractive and slender, tastefully dressed and urban-cool and carefully professional. PR is not, generally, a pretty business, and some of its practitioners can at moments slip into a certain nastiness. Belinda never did. She didn't do sweety-sweety either. When I'd picked up, she'd just said, "Listen, Sarah Jessica Parker is going to be coming out with a fragrance."

Really, I said. So you guys are doing it, huh.

"Yep. It's coming out early fall. It's called *Lovely*."

I said to her what I always say: There has to be a story. So what's the story? The usual reply is a perky "But the story is [name of celebrity or designer] is doing a fragrance!" (You say thank you and hang up politely.)

Belinda didn't. She replied, "What do you have in mind?" I thought about it for a second—I wasn't taking it all that seriously; it was, after all, a celebrity perfume, an actress—and said to Belinda, Parker's strongly associated with New York. What I'd really like to do is wander around New York with her and smell the city.

Silence on the phone. "*Smell* the city," she said. "What does that mean."

I want, I said, to smell the brick walls in the Village with her, and the tire rubber the taxis leave on the asphalt on Fourteenth Street, the subway air coming up out of the Astor Place station, and the scent of Central Park and the brackish water in the Hudson River.

Silence again.

We can drive around the city in a taxi. Or maybe we could just walk around the Village. But I want to smell New York with her.

"And talk about the perfume."

And talk about the perfume. About how she perceives smell, and so, you know, how she creative-directed her perfume. (It seemed obvious to me. Clearly it wasn't to Belinda.)

"Hm," she said briefly, "well, I don't think she's going to like that."

Is she interested in scent?

"Actually," Belinda replied forthrightly, "she's completely obsessed with it."

I didn't necessarily believe this. OK, I said, so that's a start.

"Hm," she said.

A few days later, she called me back, laughing. "She really likes it. She'll do it."

Huh. Uh—great! (What the hell do I do now?)

"I mean, we'll have to refine it," she said.

Sure, sure, I said. I was thinking, This is weird.

I proposed the story to Andy Port, my principal editor at *T, The New York Times*'s style magazine, and she was both skeptical and interested. She asked, "Do you think it's going to be a serious scent, and do you think Sarah Jessica is actually going to be involved?" I told her what Belinda had said, that she was very hands-on during the creation, very serious about it. Andy said, "We'd want an exclusive." Back to Belinda: We'd want an exclusive.

"OK, let me check, but I think we can do that. Just on the perfume, right?" Right.

I thought, How to go about this damn piece.

It wound up being a year behind the scenes with Sarah Jessica Parker and the Coty team, not only for the launch of her perfume but for an iteration of that perfume that they created together. But that is getting ahead of the story.

Je sens tressaillir en moi quelque chose qui se déplace, voudrait s'élever, quelque chose qu'on aurait désancré, à une grande profondeur; je ne sais ce que c'est, mais cela monte lentement; j'éprouve la résistance et j'entends la rumeur des distances traversées. . . . Mais quand d'un passé ancien rien ne subsiste, après la mort des êtres, après la destruction des choses seules, plus frêles mais plus vivaces, plus immatérielles, plus persistantes, plus fidèles, l'odeur et la saveur restent encore longtemps, comme des âmes, à se rappeler, à attendre, à espérer, sur la ruine de tout le reste, à porter sans fléchir, sur leur gouttelette presque impalpable, l'édifice immense du souvenir.

I feel shudder within me something that is moving, something that wants to come up, a thing at great depth that I've unanchored. I don't know what it is, but I can feel it mounting slowly. I can measure the resistance, and I can hear the echoes of distances traveled. . . . But when nothing subsists from a distant past, after the people are dead, after the things are destroyed, all alone—more frail yet more alive, more immaterial, more resilient, more faithful—the smell and taste of things endure in time, like souls reminding, waiting, hoping on the ruin of all the rest and bearing unflinchingly, in tiny and almost impalpable droplets, the immense edifice of memory.

—MARCEL PROUST, *The Remembrance of Things Past*

THE PERFECT SCENT

BECAME THE PERFUME critic of *The Times* in 2006 owing to a series of coincidences. No one was more surprised than I was.
I'd studied in China and worked in Japan and gotten a master's
in international economics and Japanese political economy, then—
credit the haphazardness of life—became a science journalist for
The Atlantic. This led me, after a chance encounter in the Gare du
Nord train station in Paris with a biophysicist and perfume genius,
to write a book called *The Emperor of Scent* about the creation of a
new, radical theory of olfaction. I'd been talking to *The New Yorker*
about possible projects—I'd proposed articles on Chinese and Indian economic development, Japanese politics—and one day they
counterproposed, to (a bit) my consternation. They were interested
in my writing a piece on the creation of a perfume. Its development, from the first instant to the launch. Behind the scenes, real
time, full access.

THE PERFECT SCENT

I'd never considered such a project. As a journalist, I was an Asianist, and I'd happened to do a book that touched on perfume; I assumed that that was finished. But OK, I said, I'd take a look.

I started going to houses. Not one of them would do it. I proposed the idea to an American designer. I had a meeting in a midtown skyscraper with the designer's PR person. "We'd love to have six thousand words in *The New Yorker*," she said straightforwardly, then after assessing me for an instant added, "but it would contradict our entire public strategy, the myth that he makes his own scents." She said no. They all turned me down—Givenchy, Estée Lauder, Kenneth Cole, Dior, Jo Malone. The Burberry PR rep, baffled, whined repeatedly into his cell phone, "I don't under*stand*, you want to watch them make a *perfume?* . . ." Then, his neurons overtaxed, he simply hung up. Chanel considered the project seriously but then, radio silence. Guerlain reacted with shocked horror; it was unthinkable. Armani passed. Ralph Lauren's PR person never even bothered to respond.

At one point someone mentioned Hermès. I dismissed the idea. The house struck me as far too constricted. Two months later, with little expectation, I took the project to Francesca Leoni, then the head of communications for Hermès in the United States. Francesca immediately said, "This is a good project; we'll do it."

And then she presented it to Paris.

I don't know everything they discussed, but I know that Jean-Claude was an advocate, that Hélène Dubrule, the company's international-marketing director, and Stéphane Wargnier, director of international communications, were cautiously favorable, and that Véronique Gautier was the primary opponent. I say this without the slightest resentment; Gautier was protecting the house and its people. It was her job. Here was some journalist, some American. She knew I spoke French—Francesca had strategically placed us together at a cocktail reception for a photography show at the Hermès boutique on Madison Avenue, and we'd begun a conversation—but she didn't know *me*. And I wanted total access, for a year. I know that in Paris they were having discussions, and more discussions, and arguments pro and con. Those in favor smoothed feathers and quietly addressed concerns and explained what was this magazine *The New Yorker*—some of them knew it, others didn't; "That's the American equivalent of *l'Express, non?*" one of them asked me once (uh, not exactly). They (once again) went over the project's concept and (once again) who I was. And with an expert touch from those in favor, we were all guided to a place where we could see it happening.

Véronique said yes.

❧

Ellena lives near the place in the South of France where, on April 7, 1947, he was born.

His family lived in Grasse. His father was a perfumer. "He had talent," Ellena would say later with affection, "but he was a dabbler." He himself had learned his craft from the craft itself, said Ellena, and from the place. As a small boy, he would leave the house at dawn with his grandmother to pick jasmine flowers. Sometimes the women who were harvesting would sit him on a wall and demand that he sing for them. He smelled the combination of jasmine—a flesh-scented flower—and sweat. Cumin smells like human sweat.

At age sixteen Ellena began working in the factory of l'Etablissement Antoine Chiris in Grasse, one of the oldest perfume houses in the world. Then at twenty-one, he left Grasse—it was 1968—for Geneva to enter his formal training to become a perfumer at the Givaudan perfume school.

The daily schedule of the students—committing to memory the smells of synthetic and natural materials, classifying scents, botany, chemistry, learning how to build a jasmine scent, a hyacinth, a rose—he found all of it rather boring. So instead he asked Givaudan master perfumer Maurice Thiboud to give him some real work to do. Thiboud entrusted him with the job of re-creating, from smell, a perfume that was on the market. (It was a common task at the time, a sort of reverse engineering, taking some Dior perfume, say, and copying it, like young artists studiously reproducing *Mona Lisas*.) Ellena did it. Thiboud gave the young man a second perfume. Ellena re-created that one. (To amuse himself, he also deconstructed it, removing materials, simplifying the scent into its elemental form.) A third, a fourth. After nine months of observing him, Thiboud told Ellena, "I'm taking you out of the school. You're going to become a junior Givaudan perfumer under me."

The first perfume he made was a small thing, of orange and patchouli, destined for the African market.

Ellena had not gone to Geneva alone. When he'd been eighteen a few days and she was still seventeen, he had met Susannah Cusak, the daughter of Irish immigrants. She had grown up in Grasse but spoke English with a quick, sharp Irish accent mixed with touches of French. Her family were artist-intellectuals. Her father, Ralph Cusak, was a painter. Her great-uncle was Samuel Beckett. Both were Irishmen who preferred French soil. "I immediately felt comfortable in this universe," Ellena said. "Susannah liked rational argument. She taught me how to structure myself." He married her, in 1967, when he was twenty.

It was she who, as he put it, gave him the virus for reading. He read Baudelaire, Laborit. His favorite was Jean Giono. "His books give a sense to life in affirming that life has no logic," said Ellena. "Like Giono, I believe in the necessity of a spirituality without religion. I don't bother God, I count on myself, and I believe in people." He read art books, and in particular he read books on painting. Whatever would feed his developing ideas of perfume. (He got a taste for painting watercolors, something he still does.)

Susannah hadn't known anyone in the perfume business. "I didn't know this world," she said. "It's part of Jean-Claude, so it's part of my life. I enjoy Jean-Claude so I enjoyed the perfume."

In 1966 his father had given Ellena, who was nineteen years old, a perfume industry magazine with an article by Edmond Roudnitska. Roudnitska was a legendary perfumer who single-handedly built much of Dior's estimable collection: *Diorama* (1949), *Diorissimo* (1956), *Eau Sauvage* (1966), and *Diorella* (1972). The piece Ellena came across was titled "Advice to a Young Perfumer." It had changed him. Years later, at age thirty, Ellena went to visit the master in a small town called Cabris, near Grasse. Roudnitska sent him away. "You smell of synthetic musks," he said. "Come back when you don't smell of anything." Ellena returned the next morning, and the two started to talk, and Ellena spent the day in awe.

Jean-Claude and Susannah had a house built in Spéracèdes, a small town of paradigmatic Côte d'Azur idyllic loveliness, and in that house—with a one-year exile in New York for his work with Givaudan—raised two children, a daughter, Céline, who became a perfumer, and a son, Hervé, an architect. The couple still live there. There was no garden by the house, so they created one. So as not to make any aesthetic mistakes, they planted only blue or white flowers. Then they added olive trees, fruit trees—cherry, apricot. Bernard Ellena, Jean-Claude's brother—also a perfumer—lives nearby. Susannah's brother lives in the house next door and has a small vineyard. Every year they harvest the grapes and make wine, which is ready by the holidays. At Christmas there are thirty of them.

⁂

Hermès had made the decision to take Ellena as the house's perfumer either very rapidly or very slowly. It depended on what day you asked him.

When the family started discussing Ellena, he was unaware of their interest. They, however, were well aware of him. He had—as an external perfumer at one of the anonymous scent makers called Symrise—just made them a scent.

The creation of a perfume begins with "the brief," the conceptual road map of the perfume that the designers and luxury houses and the creatives give the perfumers. Basically, the brief is the description of the new scent that they have in their minds. They may convey it to the perfumer in a single sentence. They may write pages. Givenchy created a brief composed of images; the concept of *Acqua di Giò* was Armani asking for the smells of Pantelleria in the Sicilian islands, where he has a home. For *J'adore*, the creatives at Parfums Dior simply told the perfumer Calice Becker to create a fragrance "as sexy as a stiletto and as comfortable as a pair of Tod's." (Becker created a multimillion-dollar hit.) The creative team responsible for the perfume *Vera Wang* saw a giant bouquet of white flowers in her store. The brief they gave the perfumer Harry Frémont was, essentially, to recreate it in a bottle. Briefs can be videotapes, songs, paintings.

Hermès's briefs were highly determined by a peculiarity of the house. Each year, Jean-Louis Dumas came up with a theme to guide the house. If Hermès launched a perfume that year, like all Hermès products it somehow followed that theme. In 2002, Dumas had chosen the Mediterranean, and Gautier, newly installed, had created her perfume brief from that. She had discovered that the Tunisian-French woman who designed the window displays in Hermès's boutiques had a garden on the beach not far from Tunis. The brief she sent out to the perfumers at the various Big Boy scent makers (among them Ellena) dictated, "Make me a perfume that smells of the scents found in this Tunisian garden." Ellena had thought and mixed things and agonized a bit and changed the mix and sent in his submission with those of his competitors. He wound up winning the brief and creating *Un Jardin en Méditerranée*. A Garden in the Mediterranean. It was Véronique Gautier's first perfume for Hermès, Ellena's second. He'd done the delectable, sparkling *Amazone* for the house in 1989.

Without his knowledge, this had put him on the family's map.

With the launch of *Un Jardin en Méditerranée* in early 2002 they started talking internally about him and the possible perfumer's position. In Paris, Ellena met Jean-Louis Dumas-Hermès, and they chatted. Ellena's wife, Susannah, was with him, and she remembers Dumas making some typically elegant comments to her: "I like your husband; he's subtle and intelligent, and it's nice to work with him." Ellena was flattered and thought, That's nice, and then thought nothing more about it. Later he heard (he doesn't remember how) that Dumas had said to Gautier, "You should go see Ellena; maybe we can do something with him," and the expression struck him. What could it mean. Probably another commission for an Hermès perfume. Which was great.

"I'd learned," he would say much later, "that everything at Hermès is slow. Which I like, because I'm slow too. I don't like fast things. I'd had a few conversations with [Dumas] of a few minutes each. The man looks you right in the eyes like a child, ready to be delighted. He poses pertinent questions, with just a little control on your points as you speak. They never, ever told me they were considering me as in-house perfumer; it simply happened like a level of oxygen rising very slowly in a room, and it's a tortuous system because you become completely seduced by them and at any moment the bottom can drop out from under you. And at the same time you're not even sure you want it. Or that they're even thinking about it. Until they tell you they are."

In February 2004, Véronique Gautier called him. Not a formal offer. Not yet. Just an idea. Very quiet. Still, she was extremely excited. *"Qu'est-ce tu en penses?"* So what do you think? He was still caught extremely surprised. "I can be sort of cold in my reactions," said Ellena, "which is to say that I don't jump around. It was interiorized."

Ellena said, We have to see each other. Gautier got on a plane with Stéphane Wargnier to the Côte d'Azur. Wargnier has huge longish curly hair and a presence as large as Gautier's; they tend to make each other expand with exclamations and observations. Wargnier always appears to have secrets and to be on the excited verge of maybe sharing them with you. Where she dresses with rich sobriety, he tends toward brilliant sapphire blue shirts and touches of exuberant Cuban reds and hot pinks mixed with expensive jackets and strange, exotic shoes. Wargnier's style is seventh-arrondissement chic with a nod to Rio de Janeiro.

Wargnier had operated at the top of the French luxury goods game for a while and was known in those circles. He had both supporters (for his control and style) and detractors (who found his particular flamboyance less than appealing). He also had, both sides acknowledged, the complete confidence of Jean-Louis Dumas-Hermès.

They met Ellena in a restaurant, La Bastide Saint Antoine. "Jacques Chibois," said Ellena (referring to the chef), and then added not entirely as an afterthought, "*deux étoiles.*" Two stars. They talked at dinner about the possibilities. He found it a grave responsibility and was cautiously elated and cautiously unnerved. To be the *parfumeur d'Hermès,* to represent Hermès. He found them very positive about this role—yes, they said, he'd be used this way, put before the public as an Hermès creator "*mais de manière trés soft.*" But very gently.

Ellena admired the house, though he wasn't a consumer of Hermès products. "*La mode ne m'intéresse pas,*" he said. Fashion doesn't interest me. (Ellena has a very precise style, about which he is fastidious, a specific equilibrium of formal and informal that could be described as Ralph Lauren in London after pheasant hunting at a corporate retreat. It sounds fussy but actually isn't at all. It's mostly the corporate retreat part. Relaxed country slacks, obviously expensive. He never wears a suit or tie but usually a blazer and always a white shirt. Years ago he decided to, as he put it, "show himself in public" in white shirts almost exclusively. "No doubt the purified aspect.") "I like luxury," said Ellena once, "although I have no use for signs of status." He considered this statement, turned it over in his head. Then he recast the proposition. "I'm not *interested* in luxury, but I'm interested in the quality of life that is led by people who are interested in luxury." This was much more precise and, thus, pleased him.

The name Ellena means "the Greek," and though as far as he knows he isn't, he certainly looks like he carries the genetics of the Aegean. He is neither tall nor short. He possesses thick, slightly wavy Mediterranean black hair, which is becoming chalked, and the confidence of a man who is conscious of being handsome. Ellena, people said to each other, never had trouble pleasing women. *Ellena n'a jamais de problème pour séduire les femmes.* Sartre once explained why he preferred the company of women: "First of all, there is the physical element. There are of course ugly women, but I prefer those who are pretty."

They drank a bottle of local white with a smokey-woody taste, and Wargnier ordered a *rouge de Loire*, much riper and fuller. To Ellena's mind, Gautier and Wargnier made it clear he'd have the right to go in whatever direction he wanted with the position of perfumer.

They didn't, according to Ellena, talk at dinner about Jean-Michel Duriez, the in-house perfumer at the house of Jean Patou, and they didn't talk about Jean-Paul Guerlain "because he wasn't really present anymore." They talked about Chanel, about Jacques Polge, but Chanel was not, they decided, the model they wanted to follow. "I know nothing of Polge himself," said Ellena. "All I know are the products, and I find them creative and reasonable. *C'est pas du délire.* It's not crazy brilliance. He is of his time. But they're *good.* They're good. What he makes, what he puts out, it's . . ." He applauded with a silent look, then said, "*Je n'ai rien à dire.*" Nothing more to say.

They asked him how he perceived Hermès. He said he found the products generous in the Mediterranean style, and pure and sophisticated in the Japanese manner. They said, smiling, "*On se retrouve.*" We've got a match. He agreed. He told them that his perfumes were constructed like that, and what he would make for them would be generous, no intrigue, no labyrinth. You had to say, "Ah, that smells good!" That's Mediterranean. And the way you created them, that had to be methodical. A perfume must be completely thought through, Ellena told them, you had to think every angle, and *then* you started building. Impeccable materials. No matter the cost. Thought applied to the most sublime materials. Wargnier ordered coffee, and they talked into the night.

At the end of April they sent the contract.

Ellena thought about all the future commissions he would not have from Gucci and Givenchy and all the other luxury houses. Then he thought about Hermès. He said yes. It was the Annunciation of the luxury world.

～

The announcement of Ellena's appointment was made by Hermès on May 5, 2004, to go into effect June 7. Everyone in Paris had a comment (New York noted it and went back to its business lunches), though since it was Paris all the comments were off the record and many were tinged, overtly or not, with venom. "It's excellent to take Jean-Claude," said one young perfumer, who cleared his throat, squinted at the sky, and added primly, "I'm almost jealous."

They were openly admiring ("They couldn't do better than Jean-Claude," the perfumer Calice Becker said, "an excellent perfumer passionate about his métier and uncompromising on materials"). They were acid ("How nice that Jean-Claude will get to do even more of his favorite thing: talking to reporters"). They were envious ("Can you imagine the *freedom*?"). They were thoughtful, analytical ("Jean-Louis was very smart about this, and you watch, they're going to start increasing market share").

The young hotshot perfumer Francis Kurkdjian commented: "For his career it's really *une belle consécration de travail*"—a beautiful acknowledgment of his work. "And a house like Hermès. Well. A true perfumer has an expertise bigger than smelling. He does everything. You think about François Coty; he decided it all, the perfume, the bottle, the ads. Jean-Claude will be able to create a true aesthetic for the house. To know their history and tell their stories in scent."

The industry discussed his putative salary in the way the French always discuss salaries: as if the KGB were listening. A huge rainmaker perfumer at the Geneva-based Big Boy Firmenich like Alberto Morillas, who landed the biggest commissions from the biggest houses, who sold tons of Firmenich's expensive captive molecules and brought in millions, must be making €300,000 a year. Surely Jean-Louis was pay-

ing at least that. It was universally agreed that Hermès's taking someone in-house was Good for the Industry. But Ellena? He was a star,
like Jacques Cavallier (who had created the lovely *Chic*, the monster hit
L'Eau d'Issey, the monster miss but utterly brilliant *Le Feu d'Issey*). Or
Kurkdjian (*Armani Mania, Le Male*). Or Becker (*J'adore, Beyond Paradise*).
And he had a star's usual partisans and critics and detractors. All this
was intensified with Ellena because he was a darling of the media, with
whom he was famous for having a *discours de parfum*. Reporters could
talk to him. He could talk back. To the degree to which this was rare, in
part it was the perfumers, who were not groomed for microphones,
and in part the paranoid, control-freak designers, whose dogma was
maintaining the official fiction that they created their own scents. They
liked perfumers to be kept in cages in dark rooms. This was why some
perfumers liked the fact that Ellena spoke.

Naturally there was also bitter commentary—vindictive jeal-
ousy is, like *beurre blanc*, a French speciality—usually punctuated,
after a careful glance over the shoulder, with the stab of a hot cigarette. "I don't think he's the best perfumer in the world," said a
competitor, "but he's one who has a thinking about perfumery. He
presents himself as the heir of Edmond Roudnitska." Yes, the competitor acknowledged, Ellena had worked under the master. A
frown, a moment's distraction while jabbing the cigarette briskly
over an ashtray. "Roudnitska's son did that thing recently. For
Frédéric Malle? You've smelled it? Yes, yes, pretty much without interest." Back to the subject: "Now Roudnitska, *he spoke* about perfume creation, and few perfumers talk about what they do. Or are
even capable of it. Jean-Claude can. So. You know." He took a drag,
exhaled a filthy cloud. "Bravo. Or whatever."

There was derision. "I don't have a big appreciation for him actually," the creator of several legendary perfumes sniffed. "His behavior is not greatly appreciated by many people." His behavior?
"Ellena has a good reputation with important people but not with
people in the perfume industry. He's a version of a celebrity chef, a

media whore, which everyone tries to become today because the world is now based on the media whereas *autrefois* the perfumer simply focused on his work and *le plan créatif*."

It was the standard critique. The Japanese may have evolved the expression "The protruding nail gets hammered down," but it is as profoundly French as pessimism. "I won't discuss Ellena," one dowager of the French industry and creator of several classic perfumes sniffed. "He's a showman."

But others took a more philosophical approach. "Grasse is a complicated tribe," said a middle-aged perfumer. "There's a real *mafia grassoise*. You need to understand, for example, that Françoise Caron is the sister of Olivier Cresp, and Françoise is also the ex-wife of Pierre Bourdon." (Caron is the creator of *Eau d'Orange Verte* for Hermès, Ungaro's *Apparition*, and Armani's *Acqua di Giò*, Cresp made *Dune Pour Homme* for Dior and Dolce & Gabbana's *Light Blue*, and Bourdon authored *Iris Poudre* for Frédéric Malle, Dior's *Dolce Vita*, and *Cool Water*, Davidoff's blockbuster. All three are among the most important perfumers in the industry.) "Henri Robert, Chanel's second in-house perfumer and author of the brilliant *Chanel 19*, is the uncle of Guy Robert, creator of *Dioressence* and *Calèche*, who is father of François Robert of *Lanvin Vetyver*. Bernard Ellena, the brother, authored many of the Benetton perfumes, and Céline Ellena, the daughter, is a perfumer as well. These are things you know if you know that tribe.

"And the mothers there! *Putain!* 'My son got a Dior commission.' 'Well, my son got the new Cerruti.' 'Mine won the latest Yves Saint Laurent and drives a BMW.' Et cetera. Grasse is a tiny little town, and the kids leave for Paris to seek their fortunes. Jean-Claude is *grassois*, and so they all know him, and when you understand that, you understand everything. Jean-Claude knows how to talk about perfume, and the press is desperate for that, and I'm sorry, but if other perfumers are jealous it's because very few perfumers can talk about perfume. 'I put jasmine in rose.' Well, OK, so what the fuck does that mean. Nothing! And someone comes and explains it, and suddenly he's a media whore? Please."

⁂

There had been, Ellena said later, two things that really made up his mind, one before he said yes, one after.

The first was something he hadn't told Gautier and Wargnier at their dinner, but he'd been turning it over in his head. "One of the perfumes I'm obsessed with is *Eau d'Hermès,* which was created by Edmond Roudnitska, with whom I worked. We must be about forty people in all of France who love it, and that's including me. I had gone on a television show and spoken about the huge admiration I have for it, the love I have for this scent. A week later I received a package from Hermès. Inside was a bottle of *Eau d'Hermès* and a note that said, very simply, 'Thank you.' And that had really registered in my head. It had marked me. I'd spoken about Guerlain's *Après l'Ondée* and a number of other scents. Not... *imagining,* of course, getting anything from it. It was a TV show! And at Hermès, someone had seen me and noted it."

The second thing was that in late May, just before the announcement, he'd gone, slightly nervous, to Pantin and approached the *hôtesse* at the reception desk. She had broken into a big smile and said to him, "*Ah, Monsieur Ellena!* We're just so thrilled you're among us now!" He found it utterly lovely. He thought, And she's the *hôtesse!* He told her, "Hermès can thank you for this welcome."

⁂

For 2005, Ellena's debut year, Dumas chose the house's annual theme: rivers.

Gautier had decided that Hermès would be creating a collection of fragrances called the *Jardins,* and so she had spent some tense hours turning it over with Hélène Dubrule, the head of marketing for Parfums Hermès. The two of them formed the core team. Gautier was more Latin and emotional, intensely certain and didactic in the most French of ways: Here was what she believed, here were the reasons, one, two, three, logically and rationally, and then, on top of that, *feeling* it and *explaining* it. She wore rich clothing, layers, and one was very conscious of her presence. Dubrule was a counterbalance, with a brisk, professional approach, systematic and analytic. She had the look, the attitude of a manager in an architectural firm in Stockholm and without sharp edges could cut through a situation or a room or a problem, give it a brief assessment, and dispatch it. If Gautier always seemed to be leaning forward into the question before them, daring it to make a move, Dubrule seemed to be coolly inclined away from it, taking its measure, all the better to pinpoint the weaknesses. She could dominate in a chosen instant, make a point that held the room and penetrated, and then let things go on until the next time that she had something to say.

They worked together hand in glove, Gautier theorizing, Dubrule directing the logistics. How, for example, to create the perfume brief out of this theme of Dumas. They put the question before themselves and paced around it, each in her way. What was the most evocative river? The *Amazon?* said Dubrule. Mm...no. The Ganges. Gautier proposed the Yangtze. The Seine? (This, they dismissed. The Seine is romantic unless you actually live in Paris, in which case it is a combination of a Disney World Main Street in summer, a sewer in winter, and year-round a Parisian version of the Long Island Expressway.)

How about the Nile. They considered this and liked it. And, said Gautier, the perfume's name: *Un Jardin sur le Nil.* That was it. Their somewhat relentlessly brief brief was done: The smell of a garden on the Nile. There you went.

Dubrule got to work and ascertained the existence of a large Victorian garden in Aswan called Kitchener. Great, perfect; they decided they'd all go, experience these smells together. Ellena would confirm his concepts, all the materials he'd put together in his imagined garden in his imagined Saharan Egypt. From that, he'd craft *Nil*.

They left Paris on Egyptian Airlines, economy class, a six-hour flight, on May 10, 2004.

The plane was a spotless new Airbus. Crisp blue seats, white interior, quietly humming. Ellena, Gautier, Dubrule, Stéphane Wargnier, and their team. Gautier had added Quentin Bertoux, a freelance photographer who often worked with Hermès. This quiet, gentle, goateed man in his forties was the documentarian, recorder of the experience, and he loved their little exploratory expedition, loved that they "left all together, like nineteenth-century scientists who would have an archaeologist, a botanist, a guide, an illustrator," said Bertoux enthusiastically. "The French and English did that a lot. They came back with paintings and statues. The statues they stole. Well," he added quickly, "we French, we only took back the sketches we made of the things we saw."

If Bertoux's job was to observe Ellena, Ellena would have loved, at that particular moment, invisibility. "When I am in the process of creation," said Ellena, "I never know when things will start. So when they say to me, 'You're coming with us up the Nile,' I find that agonizing. Very, very hard." He paused, looking intensely at nothing while he thought. "Very hard because right there, they're delimiting my space"—he made a motion with his hands in the air that described a creative space in front of him, the materials he could use to create the perfume; the materials were his ammunition—"and I experience it as a loss. It's not necessarily bad, it has to happen, but I experience it as a loss." He paused, laughed at the irony: "And as a freedom, because in that space, I can do anything." He shrugged. Here on the plane, he was scared. Would he find the thing, the scent (whatever it was)? "And when they *give* you a name like this, just hand that to you"—he made a shoving gesture—"and say, '*Bon, tu viens*' . . . and you go. It's good for you to get shaken up. Though terrifying, of course."

Ellena thought about it for a moment. "The choice imposes itself naturally," he said grandly, and then added, "although not really. Far from it." He sighed. "One has the impression that it imposes itself naturally, but in fact I'm biased by my head, my background, my history."

Véronique Gautier, on the other hand, had no such doubts. This was a woman who evinced certainty in picking up a fork. She had the ironclad opinions of a dreadnought and complete confidence in Ellena. The two were well synced professionally and, personally, connected. In 1998 as director of marketing at Cartier, she had chosen his submission for Cartier's perfume *Déclaration,* and he still appreciated her courage in taking what he'd offered, a concept of Russian tea: smoke with woods and cold, not hot, spices. (There are, says Ellena, hot spices and cold spices. Hot is cinnamon. Cold is cardamom.) And she'd taken it even though the focus group had gone *slightly* against it. So they had bonded over that. The problem was, as both were painfully aware, that this would be his first perfume as Hermès's in-house perfumer, and the scent they produced would go under an intense public scrutiny, and the stakes were much higher in just about every way. Gautier's way of dealing with it was to call him on the phone and remind him, in a Cartesian logical manner, that he had done it in the past, "and you'll do it again."

Nothing, he replied grimly, is certain.

He'd already built the perfume in his head in France. He'd gone through it step by step: Egypt, therefore heavy smells, therefore some thick jasmine, some pastel orange flowers, plus incense, just do the equation. Had to be. That was the theory. On the other hand, he couldn't help noticing that here they were on a plane to the real thing.

The Hermès group naturally checked into the legend of Aswan luxury, the Old Cataract Hotel. It is a place that summons visions of Lawrence of Arabia dining with King Fuad, or Trollope taking tea. Sullen Nubian houseboys prowled its vast cool dim white hallways, adding to the Old Cataract's faded glamour. The travelers stood on its elegant wood porches scented with Kipling, lay on the beds, exhausted from the trip, in its slightly shabby rooms. The rooms had tall, tall ceilings and equally tall French doors that you threw open to let in the Sahara, which sat, menacingly, on the eastern side of the Nile. The Sahara was a gigantic wall of sand ten stories tall. It looked like a wave suspended momentarily, ready at any instant to crash across the river and smash the hotel into splinters.

Ellena sniffed the air of the Sahara, inhaled Aswan's beautifully dry ancient air, the smell of dust and Exodus with traces of mold and car exhaust, but he found no depth, no thickness, no pastels. He smelled the wind, but it gave him nothing. He felt he couldn't get his bearings. He was so tense he didn't sleep for the first two nights. He felt, he said, like he had folded into himself.

Breakfasting from silver trays the houseboys brought—strong dark tea and thick cream, toast like roofing tiles, and rich potted fruit jams—the Hermès team looked out from their wood balconies across the Nile to pale cliffs crowned with massive dusty cakes of ancient stone, now toppled in ruins, some minor Ozymandias lit for convenient tourist viewing from 8:00 P.M. to midnight.

Gautier had also brought a film crew, who were busily making a movie to mark the creation of the perfume, though, as was painfully obvious to everyone, nothing had been created yet. Ellena felt the filmmakers' presence constantly, like an irritation on the back of the neck, conscious of the void that they were filming. He felt responsible, like a playwright who has promised actors a play while not actually being able to deliver the script. Everyone was very gentle with him, however, didn't pressure him. He appreciated that. They pretended the tension that was building didn't exist.

They got up one morning and went to Kitchener—Gautier, Dubrule, Ellena, Wargnier, the film crew, their equipment. They found surprisingly few flowers, a scattering of indigenous species already exhausted by the desert in the dry, dusty garden—was this it? They put on brave faces and started smelling. They tried a plant called *la capucine*, which has a green sort of anodyne watercress smell (Ellena ate some in front of them), and *la lantana*, which smells (rather limply) of banana and passion fruit. Kitchener was a formal, ordered garden, the kind Ellena didn't like. They weren't talking much, but it was dawning on them that perhaps someone might have done a bit more homework. Ellena smelled *fleurs d'acacia*, a tree with a soft, sweet, frangipani-like smell, but said to Gautier, "No, this is not our story." She smelled it and agreed, dismissing it in her decisive manner.

"When I'm *en brief situ*, I walk around," he said. "I look, and I watch very closely the actions of the people with me. If someone smells something several times, that means it smells good." But he saw them smell things once and move away. They went back to the Old Cataract and sat in their rooms before the tall windows and stared out at the giant cresting wave of desert.

The perfume Ellena had built in his head disintegrated and blew away, and now he had nothing.

They walked around Aswan. The sun was so violent they felt assaulted by it. Any exposed skin burned almost instantly. The heat surged over them. They went boating, and Ellena explained a bit to Bertoux about how perfumery worked, talked about ancient Egyptian perfumes, which Bertoux found interesting. "They're difficult to reproduce, you know," he said afterward, "because the formulae are not clear." Ellena watched Bertoux taking pictures, and they talked about photography. Bertoux had to place himself exactly on the edge of the Nile's water to find interesting images. *La limite entre l'eau et la terre.* The line between the water and the earth. The smell of the Nile struck him in that its water smelled more like a sea than a river. He thought that was strange.

One night they went to a Nubian restaurant, which they found magnificent, a sort of *maison d'hôte*, like going to someone's house.

Another evening, they went to the night markets, the streets full of people and light and food and noise. Ellena wanted to buy some spices in the Aswan souk to bring back to France, and a seller said,

"Smell, smell!" and held out dried lotus roots. They were extremely ugly, like the withered hands of old men cut from ancient arms and piled on the stand. He inhaled them and found they had almost no smell, but then he noticed that the seller had macerated some of the roots in simple water in a terra-cotta jar. He smelled the water, and it gave a faint scent halfway between peony and hyacinth. (The ancient Egyptians, Ellena told them, used to do this same maceration of the root in water in these same terra-cotta jars. Today, companies like Firmenich extract lotus root in stainless steel industrial canisters.)

The seller said, "Buy, buy!" so he bought some of the withered roots, though without particular excitement. He jotted it down in a small orange notebook he always carried—the lotus, its scent—and put the notebook back in his pocket.

He took the notebook out again and noted the magnolia trees, their heavy, clammy smell. Magnolias smell like lilies that have been stored, just a few hours too long, in a damp concrete basement. They also smell a bit like sperm.

He found some *jasmin sambac*, full of molecules called indoles, which smell animalic and overwhelming. Shit is full of indoles, he explained to them, and so are decomposing human bodies—the decay creates indoles; they're the molecules you smell when you smell a corpse. It's a feminine scent, the smell of death. A perfume chemist once spoke to me of "the dusty sweet rotting smell of dead bodies." Calvin Klein's *Eternity*, crafted by the legendary perfumer Sophia Grojsman of IFF, is one of the most heavily indolic perfumes around. Ironic, Ellena said to them, that they called it *Eternity*. He took the orange notebook out of his pocket again and sketched out the bottle for them. But indoles were not their story. They still had no story.

≫⊘

The cameramen were getting up at 6:00 A.M. to film, and one morning they went out on the Nile and came back and mentioned a small island they'd found: very pretty, very cool, a little island in the river. So, nothing else to do, everyone got into a felucca, an ancient four-meter Egyptian sailboat with a little motor and a swooping dagger sail like a hawk's wing, and Ellena, who loves to sail, steered them up the Nile beneath the ancient ruins under the rock cliffs. "They're big boats, so you have to have huge arm strength," he recalled.

The Nile is black, an opalescent black that, when shallow, clears to complete transparency. The party motored past the river's curves and shallows where wild grasses grew, ornamented with African kingfishers. Falcons veered from bank to bank, water buffalo soaked, nosing the muck beneath towering, soaring walls of dark yellow sand twenty stories tall that dwarfed the little boat. White herons glared at them from under date palms. The Nile was very low, and they could see the pre-dam level in dark bands in the rocks. Boats with Egyptian families cruised by, the kids shouted "Hello! Hello!" and Dubrule waved back.

They got out on the island and walked down the little street to the Nubian village. The cameraman was filming, the sound guy was miking Ellena, and the Nubians were watching it all, a bit surprised, and that was when they saw lining that street, hanging low and very dense in the large trees planted there for shade, the green mangoes.

~✎~

Green mango is one of the most astonishing smells in the world. It is, in fact, almost shockingly beautiful. To come upon the mango trees in Java or Grenada or Cebu and lean close to the fruit, an inch away, and inhale this ethereal, potent, luscious odor is like injecting a drug. It is complex yet somehow of a piece, somehow rich and fresh simultaneously, which makes it among scents excruciatingly rare. It is a smell authentically exotic and mesmerizing, at once calming and exciting, possibly more even than the ineffable smell ginger flower petals give the instant you slice them open with a fingernail. It is also heartbreakingly evanescent: The scent is exuded by the fruit only when it is on the stem. The instant you pick it, the smell begins to deteriorate with a swiftness that shows the speed at which a living thing can begin to die. Within sixty seconds, it is essentially gone.

When they returned to the Old Cataract Hotel, Ellena went quickly to his room and took out the orange notebook. Not allowing himself to think too hard about it, he scribbled down a rough formula of thirteen ingredients, some naturals, some synthetics. They were not things he'd found in Aswan. These were thirteen perfume raw materials in his lab that he would use to re-create the feelings and emotions of what he'd found in Aswan. He specified a natural essence of bitter orange to give the freshness of green mango and a synthetic grapefruit for the mango's pungent acidity and sharp freshness. (Perfumers are wary of natural grapefruit because it has a large number of sulfur atoms, which disintegrate to form malodors; the synthetic also has higher persistence, which is to say, it lasts on the skin, whereas the naturals disappear, another way in which synthetics are superior materials.) Next to each material he jotted down a guess at the percentage of the formula it would occupy. This was the rough first formula for the perfume. He referred to the perfume, as everyone in the industry does, as *le jus*. The juice.

On the island, Bertoux had been looking down at the water, then glanced over and followed Ellena's gaze upward. "Once you notice them, suddenly you see them everywhere," he said. Bertoux immediately started taking photos of mangoes. "They were small as cell phones. The mangoes were a real joy because the scent was so strong, such a beautiful smell." He paused. Frowned. Said a little warily, "An . . . *amazing* smell, really . . ."

Ellena didn't tell Dubrule and Gautier he'd created an olfactory green mango in his head. He said nothing, kept it in his head.

Ellena refined the formula for the juice in his seat on the flight from Aswan, connecting in Cairo to Paris. He frowned at the notebook in the cool quiet of the Airbus cabin, took out two ingredients, added in incense for richness and depth, then colophane, the resin you put on violin bows, for cold.

He got a flight from Paris home to Grasse. There he went to his old lab at Symrise.

At some point Hermès would be fitting him out with his own, new lab; he would have to scout out Grasse real estate, but he didn't have time to think about that now. The Symrise perfume lab was the classic configuration with a slightly industrial feel to it. It has been years since perfumers inhabited the same rooms as their raw materials. That was the nineteenth century. Today, they sit at desks in offices that, except for the stacks of blotters and the constantly shifting lines of small glass vials on the desktops, could be inhabited by lawyers or accountants. Ellena's had large glass windows looking out on Grasse's hills. The central piece of equipment in the perfumer's office is not the pipette; it is the computer, which lists the raw materials (and their all-important, secret prices) and on which they compose their formulae. These formulae are e-mailed to lab technicians down hallways who, themselves, sit before the materials, jars upon jars stacked neatly and systematically on row after row of glass and metal shelves. The materials are carefully protected from light and heat. The technicians put together the formulae and send samples of the fragrances back to the perfumers. If a formula is simple, it may be put together by a robot.

Ellena sat down at his desk, and assembled the materials—still thirteen of them—on the standard formula sheet, each material accompanied by its product code, the precise amount in milliliters, and the price per one thousand milliliters. This formula he sent electronically to his lab tech, who assembled and mixed the ingredients and brought him the resulting draft. He smelled the draft. Then he started changing it.

Of the thirteen materials, he again tossed two: a synthetic opoponax (opoponax is a resinoid, sometimes called sweet myrrh), which he'd been counting on for a resinous smell but which, as it turned out, produced mushroom; and β-ionone, a synthetic molecule he'd bet might create his green mango's gourmand fruitiness illusion but which interacted with the other materials to give apricot. He tried subbing in another extract, and it worked. This gave him his initial perfume.

Ellena thought about this for a while. Put it on his skin, smelled it, thought about it some more. What did it feel like. When he was satisfied, he started to make different iterations of the basic formula, to give the design team a choice and show possibilities. "I never submit just one," said Ellena. "With one, you cut off the dialogue. Perfumery isn't just you sitting at your desk deciding between the *touches*. It's people communicating. Talking. Tearing things apart."

He wanted ideas from different parts of Aswan. He thought of the fresh cool inky black water of the Nile and the ancient wood of the Old Cataract Hotel, and he made the first iteration a very fresh green mango with a woody angle.

On his desk in the lab he had put one of the withered lotus root hands, and he picked it up and smelled it. He quickly sketched out a mix of synthetics that gave the scent of lotus. (Unlike roses or jasmine, it is technically impossible to extract any usable perfumery material—any smell—of lotus from lotus flower or root, so every lotus scent in the world is a re-creation, a mix of molecules, naturals and synthetics, that give you that smell.) He added his created lotus scent to his green mango.

And he did a magnolia riff, remembering the magnolia trees in Egypt. Magnolia smells heavy and clammy, but it also smells of grapefruit. He had three variations on his theme.

He picked up the phone and called Gautier's and Dubrule's offices to check their schedules. He booked a flight to Paris. He took a breath and thought, Here we go.

CHAPTER 2

WHEN I ENTERED the Sarah Jessica Parker story, she and Coty had a multimillion-dollar perfume launch coming up. What I knew was the merest basics: That she—Parker—was gambling her name and her public image on a project, that they—Coty—were gambling quite a bit of capital. That it was early spring 2005, and the perfume, which had only recently been finished, was going to launch in September 2005. I was a component of the "media part," in my case its placement (they hoped) in *The New York Times.* As for my angle, I wanted, and Coty wanted, and Parker wanted, for Parker to tell me about how it was created, what it meant to her. That was the story, or at least that was part of it.

In the spring of 2004, Peter Hess, Parker's agent at Creative Artists Agency, had contacted Coty's Catherine Walsh on behalf of his client. Walsh was interested. She cautiously put her team into action. The result, half a year later, was the October 29, 2004, meeting of Coty, Parker, and Hess, at CAA's New York offices. Simply getting to that meeting—which technically didn't even seal the deal; the final mammoth two-hundred-page contract between Parker and Coty was only signed on February 1, 2005, finally locking the thing down—had taken an immense amount of work.

Walsh had made of herself a force in the industry, as large professionally as she was physically small—just five foot two. She had thick steel-and-pepper hair that she wore in a chic, wild blunt cut, and, always, bright red lipstick. Walsh found, signed, and guided designers and famous people through the complex, daunting, often scary process of building, bottling, launching, and retailing their scents. She came from a modest Irish-Catholic home and spoke with feeling about her parents and the values she'd learned: honesty, treating people well, practicality, hard work. She was single, and enjoyed it, though she kept her private life to herself. Walsh was, very refreshingly in the corporate world, generally nonaggressive and combined a quick, nicely dry intelligence with a quiet determination. If she was confident enough to show doubt, she was also quite clearly comfortable being in charge. Only her schedule took a degree of control out of her hands: She lived on planes between Paris, New York, and Los Angeles, where—the combination changed constantly—she either owned homes or kept hotel rooms on retainer. There was also her house in Telluride, where she went every chance she got to ski, and the occasional flight for a photo shoot to Ibiza or some Caribbean island.

Walsh's right hand was Carlos Timiraos, a trim handsome forty-two-year-old man with dark hair and an intense professional exterior. This he punctured occasionally with boyish enthusiasm or an idea that just had to come out. He could be openly energetic, intentionally self-contradictory in the search for the right solution in meetings ("OK, so what if we do *this?*"), self-deprecating when presenting a launch to a room full of journalists ("Obviously this isn't *my* genius, it's the brilliant work of—"), a thorough team player.

Timiraos was meticulous. He and Walsh ran the majority of Coty's brands in the United States. They operated as a brand-seeking-and-developing unit within Coty, and for their team they'd selected Chad Lavigne and Jon DiNapoli.

Lavigne, the bottle designer, was very tall and boyishly handsome and pleasantly eager, with thick college-lacrosse-player hair and cool-geek glasses, and he looked like he'd stepped down out of some Hickey Freeman billboard on Fifth Avenue. DiNapoli, the young man in charge of designing the packaging, was more reserved. Smaller, slender, and dark-eyed, he tended to watch and wait and then formulate ideas furiously behind a cool exterior. The ideas, perfectly complete, would be presented later to the team with the precision of a Rand Corporation briefing and the style of a Bergdorf window. With thick black Italian hair and creamy olive skin over sculpted cheekbones, DiNapoli looked like a sylvan faun in Dolce & Gabbana jeans.

There both are and aren't standard contracts for perfume licenses, which is to say that everything is up for negotiation. Early on, Walsh and Hess came up with "Terms Sheets" that documented their understanding on what they referred to as the "big stroke business": the length of the agreement in years, what was in play and what wasn't, payments and royalties, language governing the approval process on both sides. Essentially it was Walsh and Coty's lawyers coming up with the initiatives and sending them to Hess, who would forward them to Parker for her comment and approval, then shuttle them back again. The September 3, 2004, "Term and Renewal" Terms Sheet included, for example, the category of products that would be manufactured by Coty, careful divisions between Parker's various activities (Parker represented Garnier hair products and would, the language stipulated, continue to do so), creative control, what Parker's time commitments and obligations for publicity would be, including photo shoots and personal appearances. And so on.

Sometime in September, Parker and Hess agreed to a resolution on ten fundamental terms, and the document went back to the lawyers for the serious vetting. And Parker and Walsh were talking. Walsh wanted to do the concept prep work for the perfume, which meant asking Parker what things inspired her, what she loved, and how she envisioned her scent so that the first meeting at CAA would be productive. It was a matter, as Timiraos put it, of asking for stimulus.

DiNapoli happened to be in Paris with Walsh, so on October 1 Walsh corralled him and called Parker and put her on speaker phone, and the two of them listened as she talked. Parker described her favorite colors: She mentioned the pale pink of ballet slippers, and Walsh remembered that. Oh, and she liked gold, "but like the color on an Easter egg, with a little of the white shell showing through underneath." She loved the colors of Easter eggs, the shape of eggs. She said she had a picture of a chair that had a hardware element called a finial (a kind of ornamental knob), and she loved grosgrain ribbons—she would go into the notions stores and run her fingers over them, feeling their texture. She also loved the striped hatboxes at I. Magnin.

Parker talked about what perfume had meant to her. Her mother, she told them, had saved up to buy a bottle of perfume once a year, and Parker remembered with crystalline clarity the smell of that rare treat when she brought it home. She told them about all the strong women she'd had in her life, and how when she hugged them she smelled their perfumes and her memories of them would be created and anchored in her mind along with those scents.

Walsh called the team in New York from Paris and downloaded the conversation to them on the phone.

On October 19, Walsh and Timiraos had a one-hour conference call about Parker. It was ten days after that call that, at CAA, they had what Timiraos referred to as their big "we're all in agreement meeting": Parker, Hess, and Treciokas in the Parker camp, Walsh, Timiraos, Lavigne, and DiNapoli from the Coty side.

After more than half a year of discussions, it was the first time they all actually met.

The CAA meeting ran from 1:00 to 3:00 P.M., and DiNapoli and Timiraos had gone in early to set up. CAA called in the sushi and popcorn. Parker arrived, her people came in, they sat down in the conference room, and DiNapoli opened his bag of tricks.

He'd flown back to New York and gone out on the hunt and finally on eBay had procured a thirty-year-old I. Magnin hatbox. He set it on the table, and Parker was startled. He produced a board of grosgrain ribbons he'd put together and then a pair of pale pink ballet slippers, and she was delighted. He'd dipped eggs with Easter egg dye—which he'd eventually found, with a little difficulty, in October—in varying shades and hues and intensities of pink, one gold. When Parker saw DiNapoli's eggs she said, "Oh my god, I brought an egg too!" and surprised them by pulling a raw egg out of her purse. She dug into the purse again and brought out a crystal dish whose design she loved. They watched in surprise as she started piling objects from the purse onto the table, bits of perfumes she loved, the photo of the chair with the finial in a room with gray-and-white-striped wallpaper.

They were, they said later to each other, all struck by her incredible gratitude at their preparation for the project. To DiNapoli she said, "Your mother must be really proud of you." She said, "I can't believe you did all this for me," and they didn't know what to say. Of course they had.

Once the all-in-agreement meeting had happened, they started regular meetings at CAA. (The lawyers were still worrying the details, but they ignored this.) One time they'd get a large conference space, the next a small windowless room, hot as hell with a cramped table. They didn't care; they just worked. Parker would arrive and show them her BlackBerry, which was always with her, filled with memos she'd been generating in the middle of the night with ideas,

names for the perfume. "See?" she'd say to DiNapoli. "Three
A.M.!" She would walk into CAA very comfortably and come and
find them wherever they'd encamped their little work group, and
they'd get to it. One time the Coty team arrived late (which particularly bothered the punctual Timiraos) and Parker was sitting there
by herself in the waiting area quite contentedly, and they were a little surprised that no one was paying her any attention, but she
jumped up and said, "Great, you're here! Let's get started!"

They usually found it fun, and slightly surreal, the latter mostly
a result of the light friction of the two worlds getting to know each
other. Timiraos and DiNapoli were both highly amused by Hess.
Here was this heterosexual businessman in a suit who brokered deals
in an office and had the smooth demeanor of an ESPN anchor, and
the guy was smelling perfume and picking through pink Easter eggs.
Hess was a bit baffled by it. "I can't believe," he said at one point—
he was sitting there in his sober tie, pink all around him and a sardonic look on his face—"that we're talking about grosgrain."
Which he intentionally pronounced "gross grain."

"The business guy meets perfume!" Timiraos announced grandly
to the room, and they got some mileage out of it at his expense, but
Hess was good-natured about the whole thing. He was also, truth be
told, performing for his audience. Hess was the agent; he knew how to
play the role that allowed the parties he'd brought to the table to bond
and relax. Perfume was now his business. And the perfume industry
came close to grossing what the movie industry did. It was all part of
the talent agencies enlarging the role of celebrity, finding new ways to
brand it and monetize it.

To a meeting on November 11, Timiraos brought perfumes "as
benchmarks" for taking Parker's aesthetic temperature. This was the
beginning of the formal creation of the perfume's brief.

Timiraos had taken the names off the bottles and just labeled
them with olfactory categories. They wanted to see what she

thought. "Floral Aldehydic" was one. (It was actually *White Linen*.) "Fresh Citrus" (*Happy* by Clinique), "Green Floral" (*Lauren*), and "Contemporary Floral Bouquet" (*Romance* by Ralph Lauren). They smelled *Paris* and perfumes from Michael Kors and Marc Jacobs. Throughout, the Coty people watched her carefully, observed her face, and wrote down what she said. She didn't like many of the perfumes, and they learned quickly, and a bit to their shock, that her tastes were decidedly unconventional. She disliked florals, intensely disliked the girlish sweets and "*hated*"—she made a face that was stronger than the word—the traditional feminine constructions. She shrugged her shoulders at the aldehydes, those conventional powdery molecules, and was indifferent to pretty fresh things like lemons and the fresh-cut-grass scent of cis-3-hexenal.

What did she like? Well, first there was body odor. (They stared at her. She stared right back. Yes, body odor. "I think we all secretly really like it," she told them forthrightly, "and we're just afraid to admit it.") She liked dark scents, mustiness, slightly serpentine complex greens, labdanum, and opoponax. She liked masculine notes, and the team watched her inhale them eagerly and intently. They shot each other a few glances. They were all for avoiding the clichés, sure, and it was good—in fact it was surprising and terrific—that she was so intensely interested, but she had very, very strong opinions, and her olfactory preferences didn't accord much with her girlish, fresh-faced public image. Parker's perfume from the very start was looking like it was going to be something unexpected. The trick would be in guiding the thing.

And that was when, somewhat to their surprise, Parker showed them exactly what she wanted. She opened her purse and pulled out three different scents and set them on the table. The Coty people peered at them; none had anything to do with the others. Each was surprising, each one distinctive. Parker explained them: The first was a cheap drugstore scent, the second a very sophisticated, expensive perfume, and the third was a dubious-looking scented oil that she'd bought on the street downtown. This, she told them, was what she mixed on herself, and she put the three scents on her arm and held her arm up to them, and they smelled her fragrant, dark, musky, rich concoction. In an instant it had become very clear that Parker had a very specific scent in mind, and that moreover that scent would be quite different from anything the public would imagine from Sarah Jessica Parker. Walsh made a note to herself. She looked over at Hess, the agent, taking it all in, then at Timiraos and DiNapoli, who were digesting this.

Walsh and Timiraos wrote up Parker's dark, dirty, sexy brief, her homemade scent, and—standard practice—sent it out to the Big Boys, the raw materials houses and scent makers. Mostly this means midtown Manhattan: West Fifty-seventh Street between Ninth and Tenth avenues to IFF's strictly business offices, Fifty-seventh just west of Fifth to Givaudan's gorgeous industrial artsy spaces, Madison and Forty-third to Firmenich's pleasant, corporate-style New York headquarters, just above Bergdorf's to Fragrance Resources, and to Madison opposite the Sony building to Mane. (When they submit to Drom, the briefs travel down to Drom's so-hip-it-hurts camera-ready movie set of a space in Tribeca at 144 Duane Street. When the perfume is being done in Paris, the briefs are sent, inevitably, to Neuilly, the posh residential preserve just west of the seventeenth arrondissement where all the Big Boys have their headquarters.)

The Parker brief caused a stir. The buzz among the scent makers was that this would be a major commercial perfume; this was of course a guess, but the equation was movie star plus Catherine Walsh plus Coty, and it was promising math. Perfumers in New York and Paris examined the brief like semioticians, looking for clues, unique angles, searching for ideas, and then reached for molecules and essences and started building their proposals. When the submissions came in, Walsh and Timiraos lined them up, put them on, and smelled them. Wore them to dinner, put them on again, and showed them to people. They asked each other: "This one? That? . . ."

None of these initial juices they showed to Parker. Walsh—whose practice in this regard is standard across the industry not only with celebrities but with fashion designers as well—believes firmly that giving the non-perfume-specialist artistic director an excess of choices is a good way to create chaos and disaster. This is not caprice, nor does it involve being a control freak; between the ethyl maltol, the phenylethyl acetate, and the "Abs Foin Molecular Distillation 50 PCT BB LMR" it is amazingly easy for non-professionals to become overwhelmed and lose their way. If it cut Parker out of a step, it also streamlined her experience and removed her from a process one evaluator described to me as "the initial cluster fuck" wherein different perfumers submit different juices en masse, bottles are spilling across the table, "and you're doing triage right and left simply in order to clear the decks, find the threads, and begin."

By themselves, Walsh and Timiraos started eliminating submissions—for one, the submission of perfumer Jean-Claude Delville of Firmenich, which Delville had carefully labored over, didn't make it. Now they brought in Parker. She would, from this point, start making choices. On December 20, 2004, Walsh, Timiraos, and Parker stationed themselves in Coty's corporate offices on Fifth Avenue for the one-on-one audiences between Parker and the remaining perfumers. There were three of them.

First was Steven DeMercado from Fragrance Resources. DeMercado, an American and a guy's guy with a direct, can-do attitude, had one of the surest commercial touches in the industry. He'd done *Blush Marc Jacobs, Céline Dion,* and Calvin Klein's 1993 *Escape for Men.* He would go on to do many of Paris Hilton's perfumes. He also did Michael Jordan's eponymous scent; DeMercado, a sports fan, had put in the smell of the back of Jordan's neck.

Second was Yann Vasnier from the perfume maker Quest. Vasnier was quite young, slender, gay, and French, and he dressed in impeccable fashionable urban style. He had, in a short time, shown himself a force to be reckoned with. His strength was a startling olfactory inventiveness, and he was a metamorph, capable of crafting juices in all styles. He had created an astonishing collection of scents that illustrated olfactorily the city of Lilles—amazingly realistic smells of the paving stones of streets, of the canals—and he had two Comme des Garçons under his belt.

Third was one of perfumery's established heavyweights, Dominique Ropion from International Flavors & Fragrances. Ropion was an author of serious yet commercially successful olfactory work who also possessed a range. He could go from the ultraniche, literary *Une Fleur de Cassie* for Frédéric Malle, an exquisite and rather rarified scent, to three in Givenchy's commercial luxury collection, *Amarige, Very Irresistible,* and *Ysatis.*

In the December 20 meeting, the three perfumers listened to Parker's comments about their juices, noted her suggestions and those of Walsh and Timiraos, and then went back to their labs to prepare the next modifications. Over four months, they continually generated mods, meeting with Timiraos or Parker weeks later, then going back to make yet more mods. Early in the process IFF's Ropion, due to responsibilities on other projects, stepped aside and handed the baton to two other IFF perfumers, Laurent Le Guernec and Clément Gavarry.

Both were young. Le Guernec, in his late thirties, was the older of the two by ten years, mild-mannered, and always professional-looking in his crisp office shirts and conservative haircut. He was a Parisian who had fallen in love with an American and enthusiastically adopted New York as his home. At the time he joined the competition, Le Guernec's professional track record was scant, with few scents to his name. In 2000, he had won a Michael Kors brief and created the astonishing *Michael,* a gorgeous commercial tuberose—one of the trickiest raw materials to manipulate; he had expertly calibrated the dosages to produce one of the best perfumes it is possible to buy—and just the previous year, in 2003, Le Guernec had created no less than four perfumes for the niche New York house Bond No. 9, including *Chelsea Flowers* (whose light floral character was utterly antithetical to Parker's taste) and *So New York* (whose darkish woodiness was not).

Gavarry, whose father, in the Grasse tradition, was the famous perfumer Max Gavarry, was just as quiet, a sweet-tempered man who could be mistaken for a boy. He had rather breathtaking sea-green eyes, a Byronesque jawline, and longish thick dark wild hair and looked like a cross between a Ralph Lauren model and a Greek athlete. Gavarry's young record was similarly short, though he too had one excellent perfume under his belt, the exquisite *Carolina*, which he'd created for Carolina Herrera, a perfume that promised of its creator excellent balance and technical control. The two took over and started generating mods for Walsh and Parker.

If you are a perfumer, you generally know whether things are going your way. When your mods are met with enthusiasm and the creative team says, "You're doing great, we're loving your work," they usually mean it, although a panel test or a focus group that turns against your latest submission can reverse your fortunes in minutes; suddenly the competition looks a lot better. And as always there were the politics and the personalities. Walsh had hired two consultants—"evaluators" in industry parlance—who didn't attend the meetings but who were, the perfumers knew, smelling each submission. If two or three weeks after your last meeting you were told, "Well, we just aren't finding exactly what we'd hoped," you would probably soon be shown the door. DeMercado and Vasnier heard the message, which Timiraos delivered as gently as he could.

Gavarry and Le Guernec built the first draft of *Lovely* from Parker's musky, dirty idea. They turned it in to Walsh, Timiraos, and Parker, and that's when the discussions got intense.

Lavigne had been working on the bottle design, and it was finished by early December—he had produced a crystal bottle in the shape of an egg, with a grosgrain ribbon around it—and the packaging came together quickly after, but they were still working on the fragrance in February.

CHAPTER 3

ELLENA SPEAKS WITH such easy fluency and apparent pleasure in discussing his craft that you are always surprised at seeing him step offstage, then instantly slip away before your eyes as if the camera and lights had been turned off, allowing him to abandon the public persona. I once observed to Céline Ellena that her father divided into two, and that I was conscious of being presented with one of him while the other one watched us. "Yes," she replied evenly. "One can have the impression of knowing him without knowing him."

According to Céline, the only person who truly knows him is his wife. "My mother is the only one in whom he has confidence. They are inseparable, *un couple solaire.*" "A solar couple" is the French expression meaning that when they're not together, it's as if the sun has disappeared. "The love they have for one another is astonishing. He grew up with her; she was his confidante."

Jean-Claude and Susannah raised their children in Geneva, where he was at perfumery school, and Grasse, where he worked. "We were a very quiet family," said Céline. They did not direct the lives of their children. They set limits, Céline remembered. "How to behave in society and how to work at school, and within that they always offered us enormous freedom. They protected us. Jean-Claude never shared his problems with us. I remember him saying, 'The problems of adults are for adults to resolve.' He kept them for himself."

Céline realized very young that her father had a job unlike other fathers. "I liked that. It meant I wasn't like other children, and I had something in my father that they didn't. Once a year at school, the teacher would ask us, 'What does your father do?' And although I never got very good grades, when I answered, 'Perfumer,' the teacher instantly paid attention to me. At least one day a year I was someone truly interesting."

Susannah, to her daughter, was "always more a woman than a mother. Which suited me fine. It meant I saw her with better perspective. There is something fiercely independent in her that comes from her Irish character. She transmitted it to [Jean-Claude], an Irishness that he puts into his perfume. She protects him, and this permits him to write in freedom."

Susannah was the more tolerant one. She let her daughter *"faire l'imbécile,"* indulge in the silly joys of childhood. "When I did dress up, put on makeup, my mother always said, 'Let her do it, she's creative,' and my father found it not appropriate for a little girl. He was very serious."

But it was Jean-Claude who told them stories. Ellena would plant jasmine all over the garden, all around the house. He was also putting it everywhere in his perfumes. He cut the jasmine flowers and brought them inside, and their thick, sweet, heavy, indolic smell permeated the family. His daughter loved it, but Hervé, his nine-year-old son, found it irritating, and one day Hervé said to his father,

"OK, you're such a great perfumer, make me the smell of sweaty socks." (He thought that would be cool.) The next evening Ellena came home with a perfume, and when he opened it for his children, there before them was the smell of sweaty socks. "I actually have no idea how he did it," says Céline today. Hervé was thrilled, but his daughter was offended. "It's not a real perfume!" she said. "It's not pretty!" and her father explained to her that no, it was an odor, that there were no good and bad odors, that it was a work. It had to be respected as a work.

So Hervé asked his father for other scents, and Jean-Claude did them—the smell of madeleine cookies, of clothing, of whatever came to the boy's mind. "He didn't pull rabbits out of hats," said Céline later. "He pulled perfumes out of my brother's imagination." *Ecris-nous une histoire en odeurs,* they would say to him. Write us a story in smells. She once asked him for the smell of cloud, and he created it for her. "We asked him for bizarre things. We ordered up the smell of winter, of the snow, because we lived in the South of France, where it rarely snows."

When he got home and, with a flourish, unveiled for them this magical scent of snow, she got the idea, she said, that she could do anything, that there were indeed no limits, because here her father had gone and created a scent of a thing that had no scent. You had the story. The story of snow on the Côte d'Azur, say. You simply went and found the elements to tell your story.

"When I was a child, he didn't tell me about princesses; he told me about scents. We didn't have Cinderella. We listened to his stories in perfume."

⁘

On June 9, 2004, at 5:00 P.M., Ellena arrived for his meeting at Parfums Hermès in Pantin. He had spent the weeks since Aswan in Grasse, working on his first *essais*, his olfactory sketches, for the perfume *Un Jardin sur le Nil,* and the three small vials were in his pocket.

Pantin is an unglamorous working-class *quartier* glued to the outside ring of the Paris *périphérique* highway. The main offices of Les Ateliers Hermès are located at 12-16 rue Auger. From the outside, the Hermès ateliers look, at first, like a modest temporary project HQ, the sort of thing that might be put up for a few weeks at a construction site. Only the subtle Hermès logo of the horse and carriage under a shooting star distinguishes it. Is it disconcerting to have as one's symbol a form of transport outmoded in every way, or is it, in the country that produces the TGV super-speed train and the Airbus A380 with 500 channels of digitized video-on-demand, reassuring?

The ateliers' exterior, it turns out, is completely deceptive. Inside, the building opens into a vast, pristine, modern European corporate HQ of pale celadon metal panels, muted beiges, with glass elevators silently ascending and descending, and delicate metal bridges spanning spaces, like a collaboration between Renzo Piano and Jules Verne. Visitors to Hermès wait quietly, swallowed in the sofa and chairs. Some are models of chic. Some wear the strangely frumpy cheap suits many Frenchmen favor. Workers (always men, friendly and practical-looking) walk around in schools like fish, wearing their *bleus de travail*, the cobalt-blue overalls of French manual labor, cartloads of boxes upon boxes and bags colored the rich, powdery Hermès orange. The young women wear chic skirts made of the same material as FedEx envelopes and chic shoes with odd, lethal points on the toes. The roof is a bay of glass and metal. In the center of the lobby is a life-size wood horse with a single leather Hermès halter running its length.

Parfums Hermès is housed across the street, in an unremarkable office building, where the luxurious Hermès touch exists more or less exclusively in the sanded glass doors that adorn each office, etched with the same horse-and-carriage logo.

The meeting began forty-five minutes late due to a snafu Gautier had to resolve. Dubrule and Ellena waited in Gautier's large, light-filled modern office. Gautier walked in dressed head to toe in Hermès, and the room was instantly filled with the force of her presence.

Dubrule was as always lovely, perfectly thin with her midlength brown hair flawlessly arranged, subtle jewelry, and tasteful, tailored cottons. She gave off a British crispness. Ellena, cool as a cat, was in his uniform: sportcoat, casually rumpled button-down oxford, no tie, khakis.

Dubrule, Gautier, and Ellena arranged themselves around the conference table. I sat at one end—they helped me find an electrical outlet for my computer—and took notes. It never once bothered Ellena. He has the ego to believe that what he thinks about is important, the intelligence to make it thoughtful, and the style and skill to put it across in a concise, detailed way. He has the informality of the French, which is to say he has the mode that, in a reactionarily formal culture, acts as a facsimile of informality: Informal interaction is as carefully crafted and ornately stylized in France as its officially formal counterpart; it's simply delivered in a manner designed to give the appearance of being relaxed. But Ellena was in fact substantively more relaxed as well. He said *tu* to me pretty much from the very start without fuss. Once, he misunderstood something I'd said and, intensely disliking what he'd heard, did what people do in French, accidentally slipping back in the instant, eyes narrowed, voice icy, from the familiar *tu* to the formal *vous*. (I've done that a few times as well—suddenly I'm talking very tightly and briskly, and some verb comes out in the colder second-person plural—it's embarrassing because you inadvertently show your real feelings for that person.) We cleared up the misunderstanding, and he continued, utterly unbothered. The exception confirmed the rule; Ellena is a good communicator, in his element talking about perfume, and even when we disagreed, he engaged fully.

Hélène Dubrule also said *tu* to me almost immediately, with ease. Dubrule presents as slightly starched, but she in fact has a supple character. She manages to exercise an unblemished professional exterior, a surface as unfussily and subtly elegant as her clothing, and yet she has refreshingly little sense of hierarchy. Véronique Gautier has a much more hierarchical instinct and a highly developed sense of self and nonself. If she had said yes to letting me into Hermès, months after I'd begun attending their meetings Gautier was still saying *vous* to me in conversation. It was not unfriendly. But it was very correct. You could see the dividing lines, and you could see her eyes firmly on them.

Olivier Monteil, the head of public relations for Parfums Hermès, was at my request speaking to me with the informal *tu*. This with a palpable discomfort, at least in the beginning. Blond, blue-eyed, young but more than serious enough for his age, controlled movements in pressed cotton clothing, he was always a model of professionalism, but very careful professionalism. Whereas Jean-Claude would drop his guard with me at certain moments, in Olivier's presence I never for an instant forgot I was the journalist and that he was watching me at least as closely as I was watching them. He held his cards politely and very close to his chest. A colleague of his described Olivier to me, perceptively, as "a very private person working in public relations."

Gautier called, "Christiane!" ensuring a supply of tea and/or coffee. Anne-Lise Clément, the young blond woman who served as the *chef de groupe* (group manager) for perfumes under Dubrule, arrived with notepad and pen. They chatted. There was talk about the snafu, the marvelous weather, "*incroyable!*"

Carefully, no one mentioned the small glass spray vials Ellena was carrying.

Initial perfume submissions are, for a house's marketing people, generally complete surprises. They—the marketers and artistic directors—make up the creative teams, and their houses own the rights to produce the brands' perfumes. BPI in Paris owns the Issey Miyake and Jean Paul Gaultier licenses and creates Miyake's and

Gaultier's scents. Puig, based in Barcelona, owns Nina Ricci's and Carolina Herrera's fragrances. Estée Lauder makes Michael Kors's perfumes for him, and the Parfums Givenchy creative team is put together by LVMH, which owns its license.

The creatives gather in conference rooms in Paris and New York, staring down the long tables at the perfumers, who've come to offer the creations, juices into which they've poured years of experience and the chosen materials from their molecular arsenals (freesia accords, α-damascones, sandalwoods). The reaction can be anything. I have sat in submissions meetings where people laughed and joked and enjoyed themselves immensely, the account exec smoothly keeping the scents rolling down the line, the clients thrilled with the smells, the perfumers beaming. And I've sat in submissions meetings where the tension was like cold gelatin and people barely breathed a word and the scent sat in the air like frozen gas. Ellena has experienced having his submissions met with kisses and exclamations of joy, and he has been greeted with a president of perfumery furiously hurling his creations back at him and stating, "This is shit! Get out, Monsieur Ellena! We have nothing left to say to each other!"

The three women had seen what Ellena had smelled in Aswan—grapefruit, lotus, orange, various wood scents—but he could have done anything. He could have come to them with sixteen submissions. Or one.

Gautier: "*Bon,* so what do you have here?"

Ellena (smiling): "I say nothing. I make you smell."

Gautier: "How many things?"

With a grin (Ellena is part expert showman if he is anything), he reached into his sportcoat pocket: "A theme," he said, "in three different variations." He placed the spray vials on the polished wood. "We smelled a lot of things in Aswan," he said. He had labeled the vials *AG3, AG2, AD1.* Silently, Ellena picked up three sets of white *touches,* paper smell strips, labeled one set *AG3,* the next *AG2,* and the third *AD1,* sprayed them all. Distributed. Silence as they inhaled.

Gautier: "*Ah, j'ai faim!*" I'm hungry.

The room was silent as they held the *touches* lightly before their faces, breathing in, moving them away, shuffling them and resmelling, eyes sightless as they inhaled the molecules, olfactory signals. They glanced at each other, looking for reaction, then returned to the protoperfumes.

Gautier, decisively: "One I like. One I don't like at all."

Ellena remained unruffled. He owns a Mona Lisa smile he often wears for professional purposes. He watched patiently, his head slightly cocked, as Gautier and Dubrule did their initial walk-through of his three submissions. "I know which one you don't like," he said to Gautier.

She glanced at him, snapped the AG3 *touche* down onto the table. "I reject AG3. Very clearly." He nodded, calm. The AG3 *touche* sat on the table like a poker card. She picked up AD1, smelled it. Then reached for the third, AG2.

Dubrule was smelling AG3. "*Bonbon à la mandarine,*" she said thoughtfully, eyes focused on some middle distance. Mandarin orange candy.

Gautier: "*Bonbon alimentaire.*" Culinary sweet.

Dubrule said tentatively, "When I first smelled AD1, I thought it was green, and I got a flash of *Jardin en Méditerranée.*"

She turned to him. She was done; she wanted to know the basic concept. Gautier turned to Ellena as well. "So?" she asked. What was his basic theme for their perfume?

Ellena told them: All three of his submissions were built on green mango.

In the Nubian village that day, standing under the trees thick with unripe fruit, Ellena had inhaled the green mangoes. Egyptian mango differs from Mexican or Filipino tropical mangoes, which are sweeter and fruitier. These were pure. He looked at them—everyone was smelling the fruit now, reaching up to the green leafy branches of the elegant trees—and he saw they were jubilant, excited. Here it was, he thought. This is our story. "The most difficult sentence to write of any novel is the first," says Ellena.

It was the next-to-last day, but he had said to Gautier, "*Pour moi le voyage est terminé, je peux rentrer.*" For me the trip is finished. I can go home now. She had said, Wait a minute! The filmmakers have to film more than that! So they had stayed, walking through the dusty streets. They ate a green melon, "green, green, green," Ellena remembered, "supergreen. It had a slight taste of ethyl maltol." Ethyl maltol is the molecule you taste when you're eating cotton candy.

Now, here in Paris, they were smelling what he'd done with it. Here were three mangoes, each one inflected in some unique way, three different themes. Ellena sat back, challenging Gautier and Dubrule to figure them out.

Gautier picked up the AG3, then threw it down again with a small grimace. She picked it up yet again and said (yet again), "Well! The AG3, absolutely not." And put it down.

So what were the three variations on the mango theme? The women turned to him almost warily, raised their eyebrows.

Ellena smiled. "AD1 is a fresh mango with a wood angle. AG2 is a lotus mango."

And the AG3 that Gautier disliked?

"AG3 is a magnolia green mango," he said.

Gautier's eyebrows snapped down; she looked grim. "Ah, doesn't surprise me," she said. Then she picked up AD1, the fresh/woody-inflected scent, assessed it for an instant. "AD1 has a very mango departure," she pronounced.

Dubrule nodded, exclaimed, "*J'aime beaucoup l'AD1.*"

"I was also trying grapefruit and resin," Ellena said to Dubrule.

Gautier: "AD1 is *very* grapefruit."

They discussed the *essais*, back and forth, the various angles, notes, shades. Gautier kept picking up AG3, the magnolia variation, and then saying, "No!" and throwing it down and then picking it up again. Despite all the years Gautier and Ellena had worked together, the air was tense. Ellena sneezed, got a tissue, said, "And I put in a *lot* of incense. And there's something in them all that you'll never guess." He grinned now. The women noted the expression, then narrowed their eyes and bent over their *touches*.

"Something we talked about in Aswan," said Ellena. He was really enjoying this.

Dubrule: "*Papier russe?*" (Papyrus.)

Ellena (dismissing this): "*Non.*"

Gautier: "*Jonc de mer?*" (Sea grass.) "We discussed that."

Ellena shook his head, tantalizing. "It's a root that you eat . . ."

Silence. They were frozen over the *touches*.

Dubrule: "*La carotte?!*"

Ellena: "*Ouiiii!*"

The carrot extract was the material he'd substituted for the β-ionone synthetic that accidentally gave him apricot. It had, in his view, worked beautifully for the green mango fruity. It was in fact illogical—by itself, as a raw material, the carrot actually smelled strongly of apricot—but it is regularly the mysterious case in perfume formulae that you add one thing to, somehow, get its opposite. The interaction of materials is more instinct than science.

Gautier loved this: "*Voilà la spécialité de Jean-Claude!* Atypical things. In *Eau de Campagne* for Sisley," she said proudly to the room, "he put in tomato leaf." Tomato leaf, *vert de tomate*, is a mesmerizingly wonderful perfume ingredient, a smell at once fresh as a slender branch ripped from a tree and rich as a strong beef bouillon.

"In 1 and 2," said Gautier with pleasure, "I smell *Jardin en Méditerranée* in the fresh wood. That's exactly what I wanted, a *lien*" (a link) "between them to get the same clientele." She cocked her head. "I find AG2 a bit more complex in the foundation."

Ellena: "I put in some tarragon." He advised them, "Whenever you eat mango, put tarragon on, and the mango flavor will explode."

Dubrule was smelling AG2: "It's a bit Chinese liquor, maybe a bit too soapy."

"Because it has rose," Gautier said to her, without looking at Ellena for confirmation (he simply gazed at her, expressionless), "and rose always reads soapy." To the room she announced regally, "And now on the skin."

They rolled up sleeves as if in a vaccination clinic. Gautier took the glass vials and sprayed clouds of the three clear atomized molecular assemblages he had built. The clouds settled into shimmering vodka slicks on her arms: right wrist, right inner elbow, left wrist. She frowned, spraying. "Wow, these pumps are not so great."

She got up, walked around the room with her arms out, eyes unfocused. Ellena, Dubrule, and Clément watched her closely, not moving. Her nose moved between the three submissions on her skin. She stopped at the large glass window of her big office before the azure sky. "AG2 is best for the top notes. And AD1 I don't like." Decisively: "Too fruit." She frowned, smelling AD1. "Hm . . . we're on a candy, sort of . . . no . . ." She moved eight inches down to AG2: "Softer. More skin."

Dubrule sprayed the three on Ellena, a doctor administering shots. His white button-down sleeves rolled all the way up, he submitted to each, obediently moving his forearms beneath the vials. After she did him, Dubrule sprayed herself. Passed the vials to Clément. The women smelled him, then themselves, walking their noses between three pools of fragrance. They glanced at each other, looking for a reaction, then returned to the smell.

Dubrule: "Completely different." She meant on the paper *touches* as opposed to human skin.

Ellena: "Completely."

Dubrule: "Much less mango."

Everyone was standing now. Dubrule and Clément hovered over Ellena's arms, which he held out like Christ atop Corcovado in Rio. They sniffed him, walked their noses back and forth between the three deep pools of fragrance placed between wrist and elbow.

"It's *strange*. On skin these smell almost the reverse of how they smell on the *touche*," Dubrule said grimly. This was a serious concern, the sort of thing that caused major headaches. In classic perfumery—the great Guerlains, for example, like Aimé Guerlain's *Jicky* of 1889—clients visited the perfumers in their ateliers. Those perfumers designed their creations to be tried on skin, and they were put on skin, on the client's skin, on Guerlain's skin, on Guerlain's assistant's skin, on the skin of the other assistant the first assistant rushed off to find. And the client sat and took tea and spoke leisurely of world events and gossip, experiencing the perfume gently unfold and reveal itself as time passed, smelling its various scenes and characters and plot points. The Guerlains of that world did not use what the industry refers to, with a slight grimace, as *papier*, paper blotters. These were perfumes expressly constructed as works that incorporate time, entrances and codas, *L'Heure Bleue*'s top notes—*les notes de tête*—blossoming and dying and ceding their ostentatious place to the middle notes, *le coeur*, which developed and then after a time left the stage to *le fond*, the base notes that glowed like the coals of a great fire.

Today, the commercialization of fragrance—the Bloomingdale's perfume gauntlet or Sephora's daunting wall of cellophaned elixirs (to which one can devote only ten minutes before the next meeting begins, or the movie starts, or the baby cries)—has made paper the medium for buying fragrance. Some can take time, leisurely swimming through two hours in Saks from island of scent to island of scent, contemplating the smells. But most can't. We reach into the plexiglass tube, grab some paper, and spray, then decide in seconds, yes or no. And the molecular structure of the perfumes, their architecture itself, has had to follow. The art modifying itself to fit commerce. And so today the perfumers (not always, but increasingly) eliminate the time component—Kenneth Cole *Black* jumps fully formed from its bottle like Athena from Zeus's head, then simply fades like turning down the volume—and perfumes are constructed to smell good on paper, not on skin, which is a perversion, unless you happen to be made of paper. Perfumers take their natural materials and their synthetic molecules that might blossom beautifully on skin but that die on paper and (reluctantly) set them aside, more arrows to gather dust in their quivers. Grimly, they take the materials that smell best on blotter paper and move them to the front. Ellena thought about the problem and sifted through his materials in his head, sorting them in these terms (although he tended to be a *résistant* and damned the trend even as he recognized it).

Compounding this is the would-be client's resistance to actually trying things on, since clients don't want to "smell like five things at the same time" though there is, in fact, nothing wrong with smelling like five things at the same time in order to get to know the five to find the right one. And perfumes must be tried on for fit just as much as (more than, actually) clothing. What people wouldn't even consider demanding of a pair of jeans, or for that matter a date, they unthinkingly demand from perfume. If there is one thing one must do with a perfume, it is live with it. It is ridiculous to evaluate Guerlain's *Chamade* in five rushed seconds between drinks in midtown and the late train to Connecticut, but that is how perfumes are bought today, and it's a fair bet the next Calvin Klein will be built according to this new reality.

Gautier took Dubrule's arm, smelling her, then rolled her eyes, said to Ellena, "She turns it into *Eau Sauvage,* can you explain that!" Her nose moved over the warm skin. Decisively: "I prefer AG2 on all the skins."

Ellena, lightly: "It's AG2."

Dubrule, crisply: "It's AG2."

Everyone was smelling their own arms now.

Dubrule: "I'm finding a lot of rose in AG2. It's not bad; it just needs adjusting."

"It's the lotus," Ellena said pensively, inhaling it and looking at the ceiling. "I'll fine-tune that."

Dubrule: "It just has to be as *tenace*" tenacious, i.e., long-lasting "as *Jardin en Méditerranée.*"

They opened the far door to Gautier's office, poured into the hall like a math class at the bell, tracked people down, and sprayed AG2 on them. Philippe in the next office came into the hall like an old man walking with a cane and bellowed sclerotically, "Jean-Claude, it doesn't smell like *Equipage!* You have to redo the formula." They found this funny, and at the same time, they didn't, really; there was a faction of the Hermès family that would be happy with the unending re-creation of fragrances appropriate to France in the 1950s. Gautier and Ellena had Jean-Louis's blessing to bring Hermès perfumes into the twenty-first century, but the family was a different matter, and the family was not to be completely ignored if one was wise.

Gautier: "With *Méditerranée* we had a cocktail of citrus and fig. Here it's lotus/mango."

Ellena: "Do you remember the green melons in Aswan? That we ate with the sugar on them? I put that in."

Gautier: "AG2 is *so* much stronger." She went back into her office, picked up and smelled the *touches* again. "Do you get the aldehyde-metallic?" She handed him the *touche.*

He frowned slightly, put it under his nostrils, and inhaled. He was lost in it. Then: "Yes, on the *touche.* Not on the skin."

Gautier smelled her arms again. Nodded. "You're right."

Dubrule had gone into the office of Benoît Juillet, the finance director. He was thrilled. "Oh, this is so much more commercial!"

Gautier: "At the moment I'm not getting anything *écoeurant* (off-putting). I was nervous about that, and I don't have it. Thank God. And the sorbet—"

Ellena: "Yes, the impression of cold."

Gautier: "—very nice. *Jardin en Méditerranée*, it was so fruity! Fig milk. This one is . . . less shaded, more sunny, dryer. Not *âpre*" (bitter). "It was softer. This is *plus sec* (dryer)."

Dubrule: "I wouldn't say *sec*. I see it as juicy. You taste the water. This is more bubbly. *Jardin en Méditerranée* is not bubbly."

Gautier considered. "*Méditerranée*, crème glacée à la figue. *Nil*, sorbet. Of course, because it's finished, *Méditerranée* is more ample, large. We have to finish this." She smelled it again, declared, "It's very Jean-Claude Ellena's signature. This works perfectly for Hermès." She pursed her lips. "Will men be able to wear it?" It was an important question they had forgotten till now. Both she and Ellena were moving toward the dismissal of the archaic division of perfumes into "masculines" and "feminines," which they understood (correctly) as outmoded, a pure marketing tool concocted to give heterosexual men permission to wear fragrance. Though Gautier was, wisely, cautious of being too ahead of the market, she had decreed that the *Jardins* collection was to be unisex.

Ellena: "I was very careful not to do a cologne."

Gautier: "*Méditerranée* has something absolutely masculine that allows men to wear it. This one? Hm. We'll see. We'll have to see."

Dubrule (mildly): "Oh, I don't think there's any problem with men wearing it."

Ellena: "I tried it on numerous men; they had no trouble."

So. Scheduling. They jumped mentally to late summer, already planning the logistics for whatever AG2 would become.

Dubrule consulted some papers and reminded them that they were aiming at July 10 for final confirmation of the formula for the juice.

Gautier frowned. This deadline was exactly one month and one day away. "Why so early?"

The answer, said Dubrule, was the European regulations on toxics and allergens: The EU had a list of banned and regulated materials to which it continually added perfume ingredients to the point where the industry has started to panic. There had been fewer restrictions a mere few years earlier when they'd done *Un Jardin en Méditerranée*, and the European regs had already increased. And, added Ellena significantly, the ingredients in *Méditerranée* were simpler than those in *Nil*.

And they had another deadline, Dubrule told the team. They were going to be producing a small book about the trip to Aswan—Ellena would write it (a nod to him). It would be placed inside the *prestige*-packaged version of the perfume. By July 25 they would need to submit all of *les documents*, the various images and text for the book.

Gautier: "No, I don't agree. They print in September. You don't need to submit so early."

They discussed and decided to submit the visual designs for the *étui* (the box) first, then send the list of ingredients later; there were already IFRA regulations (International Fragrance Association) limiting the amounts of certain potentially allergenic materials that Ellena would be able to use in his juice, and now, under a new European directive that would soon be starting in 2005, all potential allergens would have to be actually printed on the box. (This had never been done before in the industry.) Assembling the list would take time, said Dubrule, since Jean-Claude was going to be modifying the materials.

Gautier reminded them that before doing the *jus* they had to start tests for the first *dérivé*—ancillary product—which in *Nil*'s case was the scented *Jardin sur le Nil* body milk. (The testing was crucial because *dérivés* from creams and lotions to soaps, shower gels, and shampoos may, and frequently do, react chemically with the perfume's formula in completely unforeseeable ways, which is why they're considered a bit of a nightmare. There is always a danger of the formula deteriorating in the ancillary's base.)

Gautier to Ellena: "You can work on the milk as of 15 July."

Clément, the perfume's project head, peered at her papers. "We gain six weeks because we're not giving the milk to the press."

Ellena: "OK, so I'll call Bernard [Bourgeois, head technician of Hermès's factory in Normandy] and tell him I'm picking up the milk next week."

Dubrule objected: "I think it's just crazy to do the milk as long as we don't have a stable formula for the juice."

Ellena shook his head. "I have a completely neutral base that poses no danger of generating differences." He asked Clément to have Bernard send him two samples of the milk.

She poised a pen over a notepad: "How much do you need?"

Ellena: "Five hundred grams."

"How about the color?" Gautier was staring at the juice's light champagne tint. "Do we need to color it?"

Dubrule looked at it. "I like the yellow. It works very nicely with the bottle."

Gautier pressed her: "You don't want us to lighten it?"

Dubrule considered it. A brief, clear movement of the head: *"Non. Non."*

Gautier took a bottle of *Eau d'Hermès,* set it next to AG2, then *Calèche,* whose juice was a much darker gold. "Perhaps you'd like it to be that?"

Dubrule looked, thought, let out a breath. *"Bof! Non, c'est bon."*

Gautier looked around the table. *"Bon! On est assez content, non?"* We're pretty happy.

～

They had sprayed AG2 on me during the meeting. On my skin it smelled almost Chinese after half an hour. I could tell it was good quality because, eventually, it just disappeared from my skin and left nothing at all. Everything good disappears from my skin. On my skin, the bad stuff, the chemical stuff, stays forever.

～

The smell of the juice was only one variable Ellena, Dubrule, and Gautier had to juggle in the equation of creating a perfume. There were many. In fact, to Dubrule and Gautier in particular—as the marketer and the company's chief strategist, they were on the line on these questions—the calculations were so complex that the smell of the thing sometimes seemed to pale in importance. Which was not, said Dubrule with a sigh, healthy. You could start, for example, with the most fundamental problem of the perfume market itself.

Perfume is the only remaining truly French-dominated international industry, and because it is French, the industry is virulently insular, pathologically paranoid, and archaically secretive. It also probably accounts for the fact that there still remain, on both sides of the Atlantic, two resolutely Gallic values, the first a kind of gentleman's agreement never to speak badly of a competitor's work in public. One hears vicious critiques of the latest Calvin or Miyake or Givenchy, but only with doors shut. (These are, generally, alternately hilarious and hair-raising.)

The other is the solemn idea of the "purity of art." This is spoken of with equal parts pride and cynicism. A French perfume industry executive said to me: "In perfumery in France, everyone comes from the sixteenth arrondissement, they're all raised with Hermès scarves, they all have degrees in poetry and commerce from some chic school, and they all think life is actually like this. They consider that what they create is great art and that because they are French the world should come on bended knee and consider itself lucky to be blessed with their creations. You talk to a French perfume executive, and it's 'I worked for Chanel, for Dior, for L'Oréal, I launched this and that perfume, and my perfumes are wonderful, fabulous, they lost five million dollars, but who cares, they're objects of art that will live forever and conform to my immortal, pure aesthetic.'

"In America people in perfumery are interested in money, money, and then money. It's 'I did this perfume, and in the first year it did X million and the second year it did Y million, and the third year it did Z million, and here's my title and here's my bonus.' If you ask them about the beauty of their perfumes, they say, 'What do you mean, "beauty of my perfumes"?'"

In France, the industry's well-remunerated positions are, like most nice things in France, reserved for those from the right *quartiers*. This is not the case in New York, the other capital of the perfume world. There, the visiting French perfume executive with the aristocratic title, summer house in Deauville, and twelfth-century family chateau in the Loire might find himself, disorientingly, meeting a twenty-seven-year-old with a Brooklyn accent and an MBA from New York University telling him (before he's even had a chance to pose his cappuccino on the conference table) that those sales figures on the new masculine needed to look a helluva lot better by next quarter.

None of this would have been a particular problem except for a tiny little inescapable fact that had been producing growing dread everywhere: The perfume industry had been in crisis for several years. The U.S. market was flat, in France it was up by an anemic amount, in Germany and Italy it was actually decreasing, and no one could figure out Asia. Between 1997 and 2004, the value of the world market grew 32.4 percent, to $25,198,700,000. In 2004, the world in which Ellena, Dubrule, and Gautier were operating, growth was slowing to a stricken 3.2 percent. This was the market into which they'd be launching Ellena's perfume *Un Jardin sur le Nil*, in which Dubrule would be trying to create some kind of traction, and whose problems Gautier would be attempting to negotiate.

And so, formally and informally, directly as a subject and indirectly by worrying such tiny details as the color of the juice, they talked with each other about what to do.

You could start (and many did) by blaming the perfumes. In effect this meant blaming the licensors who owned the brands for which they assembled the creative teams that directed the perfumers to produce juices that mostly, if you asked Ellena, really didn't merit being produced at all. The creatives came up with the briefs, the blueprints for the scents. There were the classic briefs, whose purity was always a bit of a myth (the point has never not been money), but still, Ellena remembered distinctly when the house really would say to you, "Give us the scent of a warm cloud floating in a fresh spring sky over Sicily raining titanium raindrops on a woman with emerald eyes" and so on. And one still finds aesthetics-and-concept-first briefs—for *Pure Poison* (2004), Parfums Dior posed a sphinxlike question to the perfumers: "What is it like to have something soft and something hard at the same time?" This was art.

But these briefs were disappearing. "Basically," said a French industry figure, "with today's briefs, it's 'We want something for women.' OK, which women? 'Women! All women! It should make them feel more feminine, but strong, and competent, but not too much, and it should work well in Europe, but also in the U.S., but also especially in the Asian market, and it should be new but it should be classic, and young women should love it, but older women should love it too.' If it's a French house, the brief will add, 'And it should be a great and uncompromised work of art,' and within ten years French accountants will, with the greatest secrecy, demand a cheaper reformulation to boost profits, thereby draining all the blood out of it, and if it's an American brief, it will say, 'And it should smell like that Armani thing from two years ago that did twelve million dollars in the first quarter in Europe but also like the Givenchy that sold so well in China.'" He paused and said, "American creatives create"—(quote marks with his fingers)—"by putting together focus-group data: how many Minneapolis housewives sitting in testing centers just *loved* juice RL21K. American creatives are indistinguishable from American accountants. At least they're open about it."

Christopher Brosius, the New York perfume artist, said: "I mean, come *on!* They all do these briefs, and they're the most flowery, poetic, fiery things, 'We want this! We want that! We want fantasy and newness and invention!' and then the last line always says, 'And we want it to smell like fill in the blank with the name of the current bestseller.'" He sighed loudly, said in a flat voice, "A perfumer, a woman I know, asked me about a huge international hit . . . What did I think. I said, 'Did you do it?' She said, 'Yes.' I started circumlocuting the fact that I thought it was totally mediocre, but she understood instantly. She cut me off and said, 'Look. I did six variations, the first five were extraordinary, the sixth was god-awful boring, the safest, the most conventional. And that's the one they chose.'

"That's the way it is. Not everyone. *Not* everyone, OK? But lots of people in the industry. Personally I don't have one of those twenty-five-million-dollar marketing budgets, and frankly I think they paralyze you anyway. Twenty-five million? It's like, what *is* the *deal*? They're all talking about how fabulous everything is, but they've got a look in their eyes of pure, deadened, lifeless panic. They have no vision, they have no creativity, the few brilliant people they stumble across are not allowed to do anything interesting, it's all hitched to these focus groups which I think aren't telling anyone anything, and at the end of the day most of the industry is in a very sorry state. I think customers who buy fine fragrances are increasingly bored. 'Oh, another designer launching another fragrance.' They're not being delighted; they're not tempted."

"The problem in the luxury houses," said Sabine Chabbert, ex-head of *International Cosmetic News*, "is to have believed that the *créatifs* should have the power. *C'est une erreur.* You have to have creatives. But the perfume market in France is a disaster."

There is also the fact that a customer's method of choosing someone else's smell to put on their body is a matter of consumer psychology so complex that all efforts at managing it simply crumble. The Hermès team was always conscious, said Gautier pensively, that different perfumes could have completely different sales performances not merely by continent, where many skewed hugely, but by neighborhood in a single city. "To give you an example," she said, "the city of Paris. You take three Parisian department stores: Bon Marché, Printemps, and Galeries Lafayette. Bon Marché is the Left Bank, a bit hipper, tastes that are slightly cooler, and the number one Hermès at Bon Marché is *Un Jardin en Méditerranée*. At Printemps, which is thoroughly Right Bank and the most French of the three, *24 Faubourg* is our best seller. At Galeries Lafayette, the most international, it's *Eau des Merveilles*." You could focus-group a juice to death; you still wouldn't be able to control all the variables.

And then there was something Ellena decried. They all did. The insanely escalating rate of new launches. I had a conversation with a production-side guy in New Jersey who growled about "the business today."

What about the business today?

"The juices, they just fuckin' *drop* off the list. Like stones. Used to be you could count on 'em to stick around, and every day you'd look at the list and see your juice sittin' there, and that was cash in your pocket comin' in, cha-*ching*." (As a materials maker, he makes his money in selling perfume compound.) "Today? Nobody's anniversarying fragrances anymore! Jesus!"

What does "anniversarying" mean?

He gave a rough explanation. The industry uses the term in two ways. When the marketing people say "anniversary" as a verb, they mean a tweak in presentation, revamping the box, tweaking the image, hiring a starlet, doing a miniature relaunch without doing an actual relaunch. And so boosting sales. But the production guys, who sell compound to anyone and everyone, only care about how many pounds of compound they sell. To them, a fragrance anniversaries if it sells as much as, if not more than, it did the previous year, and that, to the dismay of basically everyone in the industry, is what almost no fragrances are doing. "Used to be, long time ago, first year you'd do four million, then next year five, next six mill, and then you hang around for five to fifteen years, maybe twenty. Today you do four million the first year, then the next you do one-fifty K."

What about *L'Eau d'Issey*?

"*L'Eau d'Issey*, yeah, sure! Miyake's gotta hit. But how many between? Zilch. It's pipeline, pipeline, pipeline, push the new thing out, and you got bigger ad costs for each new one than for any scent that anniversaries. So it's worse both ways." He paused, let the air out between his teeth. "I'm tellin' ya. It's freakin' tough out there."

Angel and Shiseido's *Féminité du Bois* need, I was told by someone who should know, four weeks' maturation, then four weeks' maceration. Less, and they won't perform. One of Dior's top technical guys, the engineers who make the perfumes run, told me LVMH's financial management was pressing him to cut down the maceration time of *Eau Sauvage* from three to two weeks. It would kill the fragrance, he said.

Much of making perfume is a matter of having faith—in the product, in the gamble. The closest industry to perfume—in its protestations of artistic integrity and its manic attention to profits, in its ratio of limpid niche gems to ground-out meretricious mass-market product, in the sums of money it risks, in its mating of visions and dreams to the bottom line—is Hollywood. Even the numbers are similar: A hit perfume generally returns more or less exactly what a hit movie does, and the sums lost on failures are identical (and identically immense). You can pass on *Waterworld* or you can pass on *ET,* and the gambling is pretty much the same in the offices across the Seine in Neuilly as in the suites at the studios in the Valley. For every affirmation of belief in pure quality—"I'm very basic," Chabbert stated to me firmly in a lovely café on the place Barcelone, "I think if the perfume is good, people will buy it"— there is someone set to lose his job by not choosing the blockbuster or by producing the work of art everyone admires and no one buys. I repeated Chabbert's comment about having faith in quality to an executive at an LVMH brand. He replied instantly, and tightly, *"Joli sentiment."* Pretty sentiment. Nice idea.

Ellena believed in this sentiment, in principle. And at the same time he didn't. It depended a little bit on where and when you brought it up with him, what mood he was in, and whether the last fragrance he'd smelled left him encouraged or grimly despairing. He would say, with great conviction, "People know real beauty." And at other times (thinking of a certain recent feminine that did $10 million in one quarter postlaunch and that he considered a towering mediocrity), he would comment tersely, People are idiots, although not, of course, in those words.

If one were to quantify both beauty and faith as these two qualities are manifested in the perfume industry, one could argue that the nice idea (the belief in pure quality) is incarnated in a single crucial number, and that is the price per kilo of perfume compound or *concentrée,* the raw, pure scent goop of ingredients waiting to be dissolved in alcohol and bottled, like an oily brown brick of pure refined heroin. The designers who used to start with their visions now had their creatives begin by laying down the price for the perfumers: "We want compound at X dollars per kilo." Everyone in the industry was discussing the collapse in the quality of the materials the houses were willing to use, which was another way of saying the collapse in the price of perfume formulae. It was an open secret. It depended on whom you talked to, but generally the figure one heard was a 50 percent fall. "We really started dropping our prices with *Opium*," one veteran told me. "The *concentrée* actually wasn't expensive." *Opium* was 1978.

The dismayed perfumers talked about it incessantly. You used to easily see €230 per kilo, and then €150 per kilo, and now you got €38 to work with? It eliminated half your palette right now, the expensive stuff, the interesting stuff, and you were left with Iso-E-Super and some cheap Indian rose essence. You just had to put money into the juice. Sometimes today the mass marketers had prices of *kilos-de-concentrée* more expensive than those of the luxury brands. Gautier knew this, and so did Dubrule, but Ellena felt it in his bones and shuddered at the bargain-basement formulae. "I've been told," one American industry figure told me, "that the average price per kilo for fine fragrances is now around eighty-five dollars. I doubt that's true. I'd say in some cases it's thirty-five dollars. In some it's fifteen dollars."

Ellena would point out that when you used better ingredients, you could get away with lower dilution, so really the houses were just making up for the lack of quality by upping the concentrations, but the accountants saw the price-per-kilo figure, and that was what counted. The *concentrée* for *Chanel No. 5* was anywhere from €450 to €600 per kilo, and *Chanel 19* was supposedly a breathtaking €1,500 per kilo because of the iris root butter. Hermès's *24 Faubourg* was rumored to be around a respectable €140 due to the expensive Indian jasmine (Ellena of course went and looked them all up), but Hermès was not immune to the rumors, and they knew it. Gautier had somehow to find the right number, which meant balancing Ellena's natural desire for creative liberty against Hermès's accounting. How much Iris Naturelle at €37,000 a kilo would Bergdorf clients pay for?

Hermès discussed all of this as obsessively as any other house—if anything, more. Gautier was not going to commit suicide by letting Ellena use some stratospheric neroli essence whose per-kilo price would make every bottle a miniature financial black hole. Short of this, however, the two of them really did believe that if the perfume was good, people would buy it. And ultimately Dubrule and Gautier's strategic response was quite simple: Gain altitude.

CHAPTER 4

T HE THING ABOUT planning on hanging out with a celebrity
is that you're going to be hanging out with a celebrity. I
found it a little discomforting, and it got more so. For
one, even the initial meeting took months, and months to put
together—and kept never getting put together: We didn't have
a date.

I had a phone conversation with Ina Treciokas, Parker's publicist. Efficient, professional, a bit sharp, a bit impatient, and protective of the client in the standard Hollywood way. Would I please
send a copy of my book *The Emperor of Scent* to her for Sarah Jessica.
Sure, no problem. I sent it. We talked about SJP's availability (I
didn't know what to call her; a friend of mine who worked with her
referred to her as Sarah Jessica, but it seemed lengthy), and yes, she
was committed to doing the piece, and yes, they were going to
schedule it, but she was always on a plane or in LA or in Europe or

somewhere. We hung up. I got a call a few weeks later from Belinda Arnold, Coty's director of public relations. Would I please send a copy of *The Emperor of Scent* to Ina for Sarah Jessica. I said I'd sent one; I could hear her shrug on the phone. I sent another one. People somehow think authors have an infinite, free supply of their own books. We don't, and that was irritating.

We finally got the date—Sunday, August 14—and then came very strict marching orders: You will wait for her on the stoop of her brownstone; you will not knock on the door; when she is finished with hair and makeup she'll come out and greet you; no photography in front of Sarah Jessica's house; two hours with her and not a second more; she won't invite you in—don't ask, don't even think about it.

What really freaked me out was when I checked my e-mail and found the "call sheet." I realize (now) that this is quite normal in fashion, but I was trained as a newspaper reporter—I'd been living in Japan and started my career in the Southeast Asia bureau of *The Christian Science Monitor* writing on economics and politics—and then a magazine journalist, primarily writing on science. I never had a call sheet before.

I called the photo editor. Scott Hall turned out to be a totally nice guy. "Who's your photographer?" he asked. "Jennifer Livingston? She's good; you won't have any problems." He was very relaxed. "Shoot whatever you want; we'll make it work." You'll just edit me out if I get in the shot by accident. "Sure," he said, "or we'll just put you in with her." Seriously? So then I spent a few minutes thinking about this.

Belinda sent me a bottle of the perfume, and I started wearing it.

I figured I'd better call the photographer. I assumed she'd be more worldly in the ways of celebrities than I was. I dialed the number and reached an absolutely delightful young woman on her cell phone directing a cabbie where to drop her off on Third Avenue. Uh, so could we get together and scout out the territory? She

The New York Times

Call Sheet
T: The New York Times Style Magazine
"Sarah Jessica Parker"

Sunday, August 14
New York, NY

Time:	12 PM (Hair)
	1:30 PM (Photographer & Writer)
Location:	West Village
Subject:	Sarah Jessica Parker
	Contact: Ina Treciokas @ ID PR
	212- . . .
	Email: ina@ . . .
Photographer:	Jennifer Livingston
	Cell: 917- . . .
	Email: *jennifer@* . . .
Writer:	Chandler Burr
	Cell: 646- . . .
	Email: *chandlerburr@* . . .
Hair:	Frederic Boudet
	Contact: Rosie Creamer @ Bryan Bantry
	212- . . .
	Email: *rosie@* . . .
Photo Editor:	Scott Hall
	T: 212- . . .
	C: 347- . . .

said sure, that she'd be happy to; how about the Saturday before seeing SJP.

On Saturday I rode my bike down to Parker's house, which is on Charles Street in the West Village, and Jennifer and I met like spies, skulking around. It was sweltering. I'd gotten Sarah Jessica's address several weeks earlier from a friend who happened to live across the street from her. It seemed like everyone knew it except me. Gwyneth Paltrow's house was just down the street, he'd said, pointing it out, and there's Liv Tyler's, a block away. Jennifer and I stood in front of the imposing structure and gazed up at it. Jennifer was thin and small and cheerful and dark-eyed and cute. "So what's our plan?" she asked. I looked up and down the street like an idiot and realized I was supposed to have a plan.

I smelled SJP's stoop. It smelled like a stoop in New York in the summer, a dull powdery stone with soot. I tried to imagine getting her to bend over and smell the stoop. I smelled the tree in front of the house, but it didn't smell like anything. Jennifer gave me a look.

We walked to Bleecker Street and took a left, and I smelled the brick wall of an art gallery. It smelled of nothing. Great. New York is an empire of scents, but they tend to materialize in entirely aleatory ways and at unexpected times. You turn a corner, you're enveloped out of nowhere in an olfactory hologram of warm steamed rice, or spectral sour milk, or acrid, cloying pot, or overheated cedar mulch (the hardware store on Tenth Avenue) with a sweaty genitals angle, or some greasy unidentifiable evil smell that leaps on you, mugs you, and vanishes inexplicably in the middle of the crosswalk at Twenty-third Street and Seventh Avenue. That they are invisible makes them no less substantial. The way to experience New York's smells is on your bike because then they come at you, sequentially and strong, the plasticky chemical scent of the excessive air-conditioning in the office buildings (you enter the scent, one, two, three seconds, you exit the other end), the smell of the Gristedes grocery vegetable aisle, the 1950s scents of the lobbies of the midcentury buildings as if from a time machine, the ripely fermented rotting garbage that fell off a truck, sweetly putrifying fruit rind from the Korean bodega (the peeled detritus of a hundred smoothies in $4.95 increments of bananas and strawberries and kiwi), but you can't find them if you look for them. They find you.

I saw a flower store, Ovando, at 337 Bleecker, and we went in and looked around. We smelled the flowers, the ones with scent, the ones with none. Jennifer loved the visuals. "We could put her here!" she said. "We could put her there!" The girl in the store was scowling at us. "I'll definitely want to shoot in here," Jennifer was saying. I asked the girl if the manager was there. She was not. Huh, OK. I gave my name and *The New York Times*'s, said I'd like to bring in (pause to figure out if this is name-dropping or if I'm ensuring that things go correctly) a celebrity tomorrow, maybe, and maybe take some photos. The girl regarded me flatly. "The owner will be here tomorrow," she said. "You can ask her then."

There was a noise. The girl stepped to the back of the store, and a moment later a blond woman came out. Sandra de Ovando, the owner. Brazilian. I explained to her what we wanted to do with this celebrity. She was cautious, a bit dubious. But she'd be there tomorrow, she confirmed; yes, we could stop by.

Jennifer and I went all over the Village. We walked up to Magnolia Bakery because it was this phenomenon Sarah Jessica had helped create with *Sex and the City*, but after inhaling hopefully we agreed it actually didn't smell that much inside. Christ, a bakery without a smell. We went into the park in the V-intersection of Hudson and Bleecker Streets, and I got down on my hands and knees to smell the ingenious spongy rubber they'd laid on the ground so kids don't kill themselves, but it didn't have any smell. Oh, for Chrissake. We went down to the Hudson River and Hudson River Park, and nothing much smelled. We were sweating through our T-shirts. August in New York. We tried Perry Street, a then brand-new Jean-Georges Vongerichten place in the brand-new Richard Meier glass buildings on the West Side Highway—we thought they'd make a nice visual backdrop, and she'd like that—but the French maître d' informed us the restaurant was not opening until Monday, one day too late.

I biked home up Sixth Avenue. Herald Square was emerald and gray in front of Macy's, the sky was blue above the Empire State Building, and it was summertime.

⁂

On Sunday at 1:00 P.M. Jennifer and I meet at Sant Ambroeus, a restaurant at 259 West Fourth and Perry Streets, as we'd planned. "Don't worry!" she says brightly. "You'll do great!" She's got thousands of dollars of camera equipment with her. I'm quietly morose. I don't have a plan, or at least not a workable one. "It doesn't matter," she says, "you're going to talk about her perfume anyway." Yeah, but... "She's supposed to be very nice," says Jennifer.

Hm.

At 1:20 P.M. we leave Sant Ambroeus. We get to Charles Street. We're eight minutes early. I realize I've locked my old black bike right in front of the stoop of SJP's West Village brownstone. I say, Screw it, and we sit on her stoop at 1:25, or rather I sit and Jennifer stands and prepares her camera equipment. The street is empty, August in the Village. I realize I haven't been solicitous of Jennifer's worries. She seems pretty serene, but she's commented just enough about needing to catch SJP moving and wanting to have time with her to set her up right and also would I mind sometimes talking to her while standing back a bit? Out of the shot?

Sure.

At 1:34, SJP comes out. I'm sitting on the stoop. I think it's inevitably shocking to see a face in the flesh that you know so well as photons on a screen. You stare at it because it looks different. And, of course, very much the same. She's smaller than I expect. She smiles, and I realize it's the blue eyes that are freaking me out, deep blue, and I think that that's what I must register from the screen without realizing it. She holds out her hand to Jennifer, says, "Hi, I'm Sarah Jessica."

"I'm Jennifer Livingston," says Jennifer, beaming—Jennifer is always beaming—and adds, "I'm the photographer," and laughs because given all the equipment it's pretty obvious.

She turns to me and says, "I'm Sarah Jessica," and I say, I'm Chandl—and she says, "I *know who you are*, and I am *soooo* intimidated!"

Uh—why?

"I'm making a perfume!" she says. "And you wrote this . . . *book* on perfume!"

Have you read it?

"I'm reading it!" she says, looking up at me very seriously. She actually seems a little agitated. I'm completely flattered, and charmed, and a bit caught off guard. "It's on my book pile. You should see it. My book pile goes up to here." She raises her right hand far over her head and cups the fingers down. It's a cute exaggeration, only very slightly pretentious, since no one's pile is that high, but I appreciate the idea. I also doubt she's reading my book, but again: same thought.

She's five foot four, incredibly thin, and the guy did a great job with her hair. Spike heels, dark-blue summer dress. She's wearing light photography-ready makeup, and she looks cool and very pretty.

The first thing SJP asks me is what my sexual orientation is. What strikes me is not the question (I couldn't care less) but how delicate she is about it, and a little awkward. "What's . . ." She's smiling. "When you date . . ." I figure out what she's asking and say, Oh, I'm gay. And then: Why?

"Are you single?"

"Yeah."

She gives a little sigh. "A friend of mine—a girl—saw your photo, and she thinks you're really cute, and she's single too, but"— she shrugs—"oh well!"

We sit on the stoop, and I get out my materials. I say, I brought some things to show you.

"Oh yeah?"

When you were making your perfume, did you smell a lot of things?

"Things?"

Molecules. (Jennifer is moving around us, aiming, crouching, shooting pictures, and I'm being very cool about it and ignoring the camera. I'm such a goddamn pro.) Scent materials. I want to show you some things I really like, I tell her, I got these from a scent manufacturer called Symrise. (Symrise made me a really exquisite kit, eighty synthetics, eighty naturals in neat rows in two boxes.) Sarah Jessica takes one of the tiny vials from me, a natural absolute of algae, and she smells it with practiced, professional efficiency, cocks her head to think about it. "Brown-green," she says, which is a good description. I add, Rich ocean. "Mm, yes." Smelling it. "Ye-es. Very dark. I love this!"

I hand her a Symrise natural pepper oil. Her eyes close as she focuses on the vial under her nose. Then she opens them. Takes a second to focus. "Amazing . . ." She's shining with the beauty of this pepper.

Isn't it great? I say, excited.

"*Wonderful*," she says. We're both grinning. "Like heat inside glass," she says. She hands the pepper to Jennifer, who puts down her camera and smells it, says, "Wow," then smells it again.

SJP is fascinated. "I *love* that," she says.

I hand her another vial. She cocks her head to let the scent settle in, thinking about it. She says "soapy," which is the best descriptor of an aldehyde's scent. It's aldehyde C-12, I say (she peers at the label), the synthetic that in concert with jasmine and rose powers *Chanel No. 5*. "*Really.*" She smells it, thinking.

Can you, I ask, place the material in your memory of *Chanel No. 5*?

She shakes her head, says, "I actually don't know *No. 5* that well."

Your mom didn't wear it?

"No. When I was growing up in Ohio, my mom and dad would save up their money, and when they could afford it they'd go to Dayton to buy *White Linen* for her. It was a pretty big deal. That was her scent."

SJP's husband, Matthew Broderick, and a friend of his come out of the house. He looks tired, and she laughs and touches his face, introduces both of them. She asks him, "What time is everyone coming?"

"Six."

"You're going to be back by then, promise me!"

He nods. "We've got lots of food arriving," he says as a warning. "Like, tons."

She rolls her eyes. "We'll have leftovers."

Matthew and friend take off and I say, OK, so the plan for the afternoon is to walk around and smell your neighborhood and talk about *Lovely*.

"Cool," she says, then, "Are you hungry?" I tell her I just had red chicken curry at this awesome new Thai place. "Really! Where?" Pong Sri, Twenty-third and Sixth. No—sorry, Seventh. One of the best Thai places in Manhattan. "Will you write that down for me?" She looks around distractedly for a pen—I tell her I will; I make a mental note to e-mail it to her—and she announces very seriously, "Gosh, I have *got* to get a slice of pizza."

She gets up, dusts the stoop off her hands, ready to roll. Jennifer goes into mobile mode, somehow (I'm sort of half conscious of what she's doing) hangs all the heavy equipment around her thin shoulders. Movin' out. We go to work.

"I always, always, always thought about creating my own scent," SJP says as we walk down the sidewalk of Charles Street toward the Hudson River. "Finally, after twenty years of having it in my head, I got brave enough to talk to my agent, Peter Hess, about it." (She gives me a sideways look. "It's funny," she whispers, "I'm still sort of embarrassed, you know? Like I really know what I'm doing, right?") "So

Peter said great, good to know, I'll get back to you. And I thought, OK..." She makes an "Uh—*right*" look with her eyes.

We walk.

"And then suddenly he called, and we started meeting companies"—potential licensors who would take the financial risk, develop, and distribute her perfume—"and it didn't click and didn't click, and then the instant I met Catherine, and, I mean, we'd barely exchanged pleasantries, and"—she looks awed—"it felt so *right* with her, and, well, I hoped she felt the same."

Walsh did. Walsh got to work, and an immense amount of internal discussion, focus groups on the impact of Parker's name and image, and numbers crunching in various perfume markets around the world led to the meeting in Hess's office. Parker talks about her with open awe. "I was *mentored* by her, in knowing fragrance and in knowing the consumer. She's mighty and powerful and a truly elegant woman. One of the last who wears really red lipstick and nothing else. It's so strong. She's tiny, I think even shorter than I am."

She looks at me. "The idea that you like something can lead you to the idea that you know something about it." She raises her eyebrows, looking in that instant both pretty and a bit alarmed. "Which is, of course, not necessarily the case."

We turn left at Bleecker. "When Catherine and I started talking about creating a perfume," she says as we walk, "I gave her my idea."

You already had a specific scent in mind?

"Oh *yeah. Very* specific because I had—in a sense—and this sounds strange perhaps, but—I had already created a fragrance, something I wore for years. Three scents I mixed on my skin, and honestly it was terrific. The grips and the cameramen would say, 'Wow, what are you wearing?' It was really successful."

What was it?

She hesitates. "Do you think it's bad to say?"

No—why?

She considers. "Well, first, I'd buy a drugstore musk, $6.99 a bottle at Thriftys."

What was it called?

"Uhhh . . ." She touches my arm, looks pretty torn up. "Do I say?" Winces. "I think I better not say."

I pretend to look hurt, she's apologetic, I cave, and she laughs. The second?

"Second was an Egyptian oil from an African American gentleman who used to sell them on lower Broadway. And, third, a fairly costly male scent." She tells me what it is, off the record. I'm surprised. It's from an edgy scent collection with a dark, very downtown, not infrequently forbidding aesthetic. It's not, I say, a scent I'd have imagined you creating.

"Oh, *it's me*," she says. "*Love* it. Really *dirty*. *Really* dirty, really sexy. I said to Catherine, 'There's nothing on the market for women like this.' "

So you *did* know something about creating a perfume, I say.

"Well," she says, "I thought I did."

We cross West Tenth.

Generally New Yorkers live up to their reputation for carefully, indeed stoically, ignoring famous people. The New Yorkers ignore Sarah Jessica. The tourists break into huge surprised smiles. "Whoa!" They nudge each other.

What scent do you wear? I ask her.

"Sarah," says a woman, so excited, "can I have a picture with you?"

I stop, though I'm not sure if I'm supposed to, and so she stops, and stopped, is gracious and smiling. "Sure!"

They take the photo.

We continue down the sidewalk. I ask, What scent?

"Guerlain's *Vetiver*."

Whoa, I say, you're hard-core! A classic masculine! Seriously?

"I love *Vetiver*!"

She stops and stares at a young couple in front of her, who are at the same time realizing who she is and staring back. "Hey, are you from Cincinnati?" she asks them. He's wearing a Reds T-shirt.

"Yeah," he says. He's blond, maybe twenty-two, beefy, smiling, relaxed; she, on the other hand, is quite startled.

"What part?"

"Cincinnati," he says, and Parker says immediately, insistent, "Yes, yes, what *part*?" and then, "I'm from Clifton!"

"Huh!" he says. "Right near us!"

She prepares to move off and by way of apology rolls her eyes significantly in my direction. ("I'm working!") They look over at me (I'm writing down "Clifton"), make an "Oh" look (a reporter . . .), and nod back confidentially (*got it*), and we move down Bleecker.

She tells me how Walsh assembled her creative group, the people who would take her ideas and turn them into *Lovely*. "I called them my mighty team, small and terrific. Jon DiNapoli, Carlos Timiraos, and Leslie Oglesby, global marketing director, who translated and educated me, and Chad Lavigne, our great bottle designer. That was it. Coty's huge, they just bought Unilever's entire fragrance division, but working with them felt totally intimate. *Sex and the City*—you know how most TV shows have, like, twenty writers?—we had *six*. It was the same feel. It was boutique."

Walsh and Timiraos put together Parker's brief. She was, she says, intimidated by the process, struggling to get out her thoughts of this scent. "I said to Catherine, 'I don't know chemistry! I don't know the vernacular!' and Catherine said, 'Yes you do; you have language.' Sometimes I grasped for words, but I'd find them, or they'd find them for me. The language was so important. So I gave my idea to Catherine—"

This idea is those three you told me about? I ask. Or didn't tell me about, the cheap drugstore—?

"—drugstore musk, uh huh, and the—"

The Egyptian oil, right? And the—

"—yeah, the masculine sce—wait, you didn't write those down, did you!"

I didn't write them down! You told me not to! (I hold up my notebook to her, and she peers at it. Didn't write it down.)

"So Laurent [Le Guernec] and Clément [Gavarry] took these scents I'd given them and everything we'd talked about, and they went away and came back with our first draft." She takes a deep breath. "And guess what," she says, very grimly. "The oiliness bothered me. And also guess what. My dirty instinct really made Catherine uncomfortable. She said, 'Listen, you simply can't sell this to a girl. The market just won't follow you there. Not yet.' *And* I wanted my scent to be genderless—not masculine, not feminine, just the scent—but they pointed out that my first time out, a feminine is so much more classic." (Which is to say that feminines are much more commercially successful than mixed scents, at least thus far.) "And I had this idea I loved, and then I didn't have it, and I thought, My God, so what do I do *now?*"

We're almost to Ovando, and I say to her, OK, so I want to take you to this flower store.

"There's a flower store near here?"

I point at 337 Bleecker. She looks at it. "*Huh,*" she says, "I don't think I've ever been in here. Isn't that strange?"

Exquisite flowers, chic black walls, the fresh, fresh stems in the air-conditioned cold, and as the stem-filled air hits us, "That smell," Parker says, "that *green.*" The girl behind the counter sees her: "Oh!" Bursts into a huge smile. Sarah Jessica smiles back, says, "Could I drop these . . . ?" (Her purse and glasses, on the counter.) The girl says, "Ahh—anything you want!" which I think is a little weirdly servile. I look at Sarah Jessica because I want to see how she'll react, and she just swims right over the remark, moving to the flowers.

We lean over some small, white, round buds. I frown. I say, Jesus, these flowers smell exactly like the algae absolute! Sarah Jessica leans in, bursts out laughing. "My God, they *do.*" She leans down eagerly over some dark green mint. Inhales it. "Mmmm! We grow mint at the beach. I pretty much don't drink, but when friends are over we crush this in drinks, put in ice, and you don't need anything else."

Sandra de Ovando comes in. Sarah Jessica asks her what the algae-smelling buds are. St.-John's-wort. "That's why we feel so good!" she says to me, and jokes, "You can get off your Prozac."

Paxil, I say, looking at the St.-John's-wort.

"There ya go!" She turns to Sandra to negotiate some on my behalf: "How much?"

"Three fifty a stem," says Sandra, and Sarah Jessica counts the stems—"three fifty, seven, ten fifty . . ."—turns to me, "Fifty dollars a month, you'll feel great, *and* they're beautiful."

I say I'll certainly consider it. We move on to the scentless white hydrangeas—"These we grow at home," she says. We bury our noses in a big bouquet of chamomile, gorgeous scents of fresh and cold green. Sandra brings in a bowl of the most sumptuous apricot-colored tea rose, and Sarah Jessica shoves them to my nose. "*These* are *wonderful!*" The scent is amazing. Tomato, I say.

"Artichoke," she says, and I'm irritated that she's hit it closer. "God, roses that smell of *rose.* You can't find that anymore. They're all deli roses now." She looks daintily grim. "Deli roses just don't cut it." She runs back to smell the chamomile again, her hair flying, Jennifer shooting her as she moves. (Jennifer turns around, mouths to me, This is *so fun.* Jennifer is shooting like a determined sniper, then pauses. "Do you want to put on lipstick?" she asks. "Oh," says Sarah Jessica, "I don't wear it.") She picks up thistles, inhales, says to me, "Chandler, you've gotta smell this!" Thistles smell of clean dust and hay. "We're around hay a lot in Ireland," she says. "We help them unload it, and my husband smells of it."

We walk a few doors down to Goodfellas Pizza at Christopher Street. She orders a slice of pepperoni and a root beer and, for me, a slice of mushroom and an iced tea. "Hey!" one of the pizza guys

THE PERFECT SCENT

says to her, "you gotta perfume now, right?" "I *do*," she says deliciously. He narrows his eyes. "Yeah?" (And she better be straight with him here.) "So—whaddya put in that stuff?" "It has a *tiny, teeny* little bit of orange blossom," she says, "and we cut in a little lavender. And patchouli."

The guy looks impressed: *This* is a woman who knows how to make a perfume. When she tries to pay, he waves it away. "Onna house." "*C'mon*," she says, then laughs and leaves a tip twice as big as the check. I'm staring at the friendly pizza guy, dark hair, wearing a sweaty wifebeater-T in front of his oven. I'm thinking, How the hell does this guy know she has a perfume? She sees my look and takes it for something else, leans toward me. "If we could all take a truth pill, us Americans, with our antibacterial soap and our deodorants—we love BO! We love the smell of *us*. Our bodies. Yes, we want to be clean, but really I think we like what we smell like."

And that's what you wanted in your perfume.

"That's what I wanted."

We're walking up toward Greenwich Avenue. The hot August sky has become dark and menacing, and we have no umbrellas. "So we worked on it," she says, "and worked on it and worked on it and worked on it. We had all our meetings in Peter's offices at CAA. Toward the end, Catherine said distinguishing our successive versions was like splitting atoms, that's the kind of fine-tuning we did."

Walsh, in fact, made an unusual decision; often, celebrities and designers never once lay eyes on the perfumers who build their scents, never speak with them or see them, never know their names. But Walsh, quite intelligently and rather daringly, decided that Parker should personally direct the final stages. As Walsh said to her, "You have the images in your head and the words you need to express them." Which is why Parker wound up sitting down with Gavarry and Le Guernec.

"I told them I wanted my fragrance to have social skills. You know how when you hug someone because you have to or because they force themselves on you? And how their scent stays on you. Usually some scent-of-the-moment, talking too loud. *Never.* I wanted mine to have manners, beauty, subtlety. My mother never let us listen to commercial radio, only the NPR station, with classical after that, and then *All Things Considered.*" She intones, " 'Eighty-nine point nine, WQXR.' It was great for everyone, kids who don't have TV, for adults, a *feeling* of culture, of calm, civilized." She puts her hand over her mouth and lowers her voice. " 'That was Fauré, Opus 26.' "

She starts to tell me about how they went about the pink dress, designed by Oscar de la Renta for the ad, the thousand other details. "What upset me a little when they premiered *Lovely* for the press," she says, "was when they'd write about my scent with others that they call 'celebrity' which"—she laughs, *very* briefly—"is not a word I'd apply to myself. I was clumped in with everyone, and it simply makes my perfume seem . . ." She searches for the word, then delicately makes the point differently. "I was at every single meeting. Every one. We worked like crazy. I brought in pieces of fabric for them that I'd worn, synthetics, natural fibers, with the latest version of the scent sprayed on them. I think people like walking into their closet and smelling their scent. I wanted to know how it smelled on different clothes.

"We tried this grosgrain ribbed ribbon around the neck of the bottle, and that one, and *that* one, we blew up my name, reduced it, changed the font a thousand times. And the scent . . . we agonized over it, contemplated it, altered it a million ways. It was so important to me because I know how my mom saved to buy her one bottle of perfume a year, two if she was lucky. Most people do not have huge disposable incomes, and I don't care if I don't make any money on this, honestly, I didn't do this to become superrich. I wanted to create this beautiful thing, and it just *has* to be worth it when people spend their money. It's that instant when you smell it, and, you know, you think, Yes. We'd be on conference calls, Catherine in Paris, Jon in New York, me who knows where, we'd change it, and I'd get the next sample three days later, put it on, put it on my sister, my clothes, my friends."

Since we're here in the Village, I figure we should talk about the smells of the Village, so I bring it up. The chalky, warm scent of hot brick, New York's perfume of warm asphalt, the hot dry cement. She crinkles her nose and smells. As she's doing this, a young guy walks by with his very young daughters and says, "Sarah?" and she almost shouts and gives him a huge hug and talks to his daughters.

We walk past Magnolia Bakery and talk about the perfumes she likes. "One I *loooooove*," she says, "it's Yves Saint Laurent, it's called—oh *God*, I've got a bottle upstairs, pink? Bottle like a jewel?"

I say, *Baby Doll*?

"No, no, the—*what is*—I smelled it the first time because I inherited an unopened bottle from a friend's deceased mother. Weird, huh. The black top?"

Oh, *Paris*?

"*Paris*! I *love* that one." Jennifer is shooting away, crouching down, darting in front of people. It's a weird thing I've noticed photographers do, honestly just not giving a damn about anyone else while they're taking shots. People sort of jump out of her way. We stop in the antiques store Venfield, and we smell the candles burning there, from Paris's Hôtel Costes, the scent of a dark, rich European library, leather and spices with a bit of cigar ash dropped on the thick wool carpets.

"This is great," she says. As soon as we step outside, huge raindrops start a slow barrage, and we run, she adeptly in heels, past the new Bond No. 9 store and Marc Jacobs and up her street. "The scent when the cement gets wet," I yell, and she shouts, "And the wet heat smell that comes up." She runs, laughing, to her house, she pushes open the door for us as the sky opens and it becomes monsoon. Her son, in pajamas and with his nanny, is surprised and delighted to see her in the entry. "What's your name!" she asks, shaking off a little rain. He thinks about it, states: "James Wilke Broderick!" James has *not* taken his nap and is no worse for wear.

We take James and go up to the living room, large and clean and cool. I look at a pile of books on top of the mantel over the fireplace in the comfortable study in back. She glances over. "See? That's the pile." She raises her right hand far over her head and cups the fingers down, and it exactly fits the pile, which is just that high. She pulls my book out—it's the sixth down of about fifteen—and hands it to me. As she walks away, I see a bookmark in the middle.

We're talking about how she feels about scent. Jennifer interrupts briefly to ask her to pose on the stairs, no, the other way, put your arms up. I'm interested to see the degree to which Jennifer is directive and Sarah Jessica is malleable, she does exactly what Jennifer asks and doesn't stop talking to me about body odor, this mystery scent she won't tell me about, as her body moves to Jennifer's repositioning her arms. Sarah Jessica says to Jennifer, "I just really don't want to look into the camera, OK? I *haaaaate* that." I'm confused. I say, You hate looking into the camera? "Ech," she says and shudders. "We never do it on the movie set."

James is still in the living room, singing a song to himself. He's being very precise with the words.

I say, So this scent you created—that you didn't use—

"—*this* time."

This time. It represents your . . . (I'm stuck.)

She cocks her head and assesses me for a moment, eyes narrowed. She says, "Wait." Disappears. Jennifer and I look at each other. She reappears. She's holding three different bottles in her hand. "Bonnie Belle Skin Musk," she says, hands it to me. The bottle alone could make you laugh out loud: cheap, tacky American drugstore perfume, tacky green cap, so evocative and fun. "Bonnie Belle called me because I'd been buying this in bulk from them," she says, sitting cross-legged, "and it's inexpensive, so the stuff disappears from your skin fast. They said, 'We've been bought, and we can tell you that the formula is going to change, so would you like to buy what's left?' For a day I tried to calculate how long I was going to live. And then I bought every single bottle they had. About five hundred bottles. I have them in storage. Here." She sees me trying to be sparing with it. "Oh, please, I still have plenty left." She stops suddenly, realizing what James is softly singing: "And we lived," he sings.

"A life of ease," she sings, filling in the next words.

"In our yellow," he sings back to her, completing the line.

Together they sing, "Submarine."

She hands me a vial. The Egyptian oil. "I have a huge jar. Try it." I apply it on top of the Bonnie Belle. The third is Comme des Garçons *Incense Avignon,* $150 for a 50 ml bottle. Smoky, heavy, dark, perverse, slightly brutal. I put that on the other two, and the three together are powerful and strange, the smell just as she'd described it on the stoop. So this is what she had in her head. I smell it.

"My next scent will be genderless," she says. "Less like an alto, more like a bass, less a pinot grigio than a cabernet. Fuller." She inhales the crepuscular incense musk mix on my arm, sits back. *"Riskier."*

Which shows exactly what is most interesting about the structure of *Lovely.* It is, in fact, a risk, and I suspect Parker doesn't fully realize how successfully that risk has been negotiated by Clément Gavarry and Laurent Le Guernec, her perfumers, and her Coty marketing team. *Lovely* is a piece of extremely interesting technical work, but it is a tightrope walk, or more precisely a deft piece of multipart perfumery. It is, in its most simple and immediate incarnation, instantly legible, placeable.

There is the classic French school of perfumery—Hermès's *Calèche,* Lanvin's *Arpège*—a philosophy that holds that the point is artistry and the ego of the creator, and the focus is the perfume itself, and when you smell these perfumes you say, "She's wearing a great fragrance." Though it's actually the fragrance that wears you; the person is merely a transport vehicle for the perfume. And then there are the modernist works of scent art, done so well by Fresh, for example, in which a material—a pear, a cup of sake, a peel of tree bark—is transformed into an abstracted scent that exists to enhance *you*. You say, "*She* smells amazing" as if (this is the point) the perfume emanated from and was part of her own body.

Lovely is an interesting fusion of both. It is French school in style, in that one doesn't "smell of" *Lovely*. One *wears* it. One puts it on. *Lovely* is the lightest olfactory party dress of powder and sweet, the scent equivalent of the terrific wrap of soft floating mesh fabric I saw one summer enveloping the shoulders of a young woman, a physical cloud she wore elegantly through the East Village streets. One notices that lovely wrap. But *Lovely* is modernist school in behavior. The perfume melts into you, and there is a point in its development when the other person will stop seeing the wrap—where the scent stops behaving like a coat—and sees only the wearer, who is somehow prettier, more delicate—enhanced—in an indefinable way. The perfume, as "the perfume," has disappeared, leaving only you.

This is, in part, why it takes a bit of time to notice that Parker has in a real sense gotten what she wanted, though Walsh allowed it in a very astute way. By the time I met her, I'd been wearing *Lovely* for a week. The scent does not, even well into the dry-down, bifurcate at all. It reveals none of its moving parts. What it does do, for those who pay attention, is reveal its structure, a sheath of light built around a core of dark. It reveals the scent of the skin of the shoulders below, the scent of a clean, warm, very human body that might be walking energetically up Bleecker Street past Goodfellas and Ovando toward Greenwich Avenue.

I doubt Sarah Jessica knew the perfumery term (*animalic;* I forgot to ask her about it), but she had the concept, and she found a way to express it. Taking Coty's probably wise marketing advice, she and the team created a lilting perfume welded to an invisible platform as masculine as it is feminine, animalic, hard-core, ever so slightly sweaty. The girl we see and the girl we don't. At first.

Do you, I ask her, regret the choice not to do the full-on dirty sexy thing? She looks at me, smiling broadly, slowly shakes her head no. "When I was thirteen, Brooke Shields and I were really good friends, and she gave me a bottle of *Joy*. I was, like, holy moley, *Joy!* My mom took it away from me." Did you resent that? "Not at all. She was completely right. It would've simply been inappropriate for a thirteen-year-old girl weighing thirty-eight pounds to be wearing *Joy*. That was a choice. With mine, I made choices."

She pushes back a strand of hair. She looks pensive for a second. "Catherine said, 'Go farther with your next one.' You know how when you travel in Europe and everything is so well marked? This destination, then that one, and the final one in big letters? I can see my next destination. I can see it, I know the name, the shape of the bottle, the ad campaign. I just can't think about it right now."

Lovely, which Coty had debuted privately to the press several months before, officially launched to the public in late August 2005. In all, Coty would spend millions on it, as would its competitors on any similarly high-profile perfume, and everywhere there was Parker's image, carefully directed and engineered by the marketing and branding visionary Trey Laird, the surprisingly modest, low-key head of Laird+Partners, which had handled Parker for her multimillion-dollar Gap campaign; Parker revered him. The image was Parker in a pale pink vaguely ballet-style dress designed by Oscar de la Renta. The pink of DiNapoli's eggs.

The perfume started rising up the charts.

I spoke one afternoon with Walsh in her corner office at Coty's then headquarters in the Trump Tower on Fifth Avenue. She was wearing her signature really red lipstick, blunt hairstyle of undyed gray standing up from her head. She was characteristically direct and strong without being aggressive and, as do most executives, took care to express things with precision.

"We met with Sarah Jessica for the first time in the CAA offices of her agent, Peter Hess, in New York. Ina Treciokas, her publicist, was also there. White conference room, very slick, very agent. There was sushi on platters—we always have sushi. We'd negotiated a certain amount before, obviously, the parameters of the deal and so on, but this was the first time I'd met her. She was dressed very casually, jeans, T-shirt—it was May—and I was talking on my phone"—Walsh took out her impossibly tiny red cell phone—"and I didn't hear her walk in, and I turn around and hang up fast, and the first thing she does is hands to mouth, 'Oh! I'm sorry! I didn't mean to make you cut your call off!' and then, 'Is that *real*?' Because she and Matthew apparently have a competition about who has the latest gadget. I was at the Chateau Marmont and it blew off the table, that's how light the phone is.

"It's not just that she's friendly, which I expected, and humble, which was a nice surprise. It's that she wanted to listen, so intently, to every word I said. I was in that office explaining how one goes about creating a fragrance and what would be the steps between now, our first meeting, and setting the first jar of Sarah Jessica's perfume on the first counter. She was so interested in hearing every word, looking right in my eyes, and she was taking it seriously. No, more: She'd done her homework. She was on. It was passion. What continues to impress me about her is how big her mind is. In my career you work with a lot of smart people, passionate people, creative people—she wraps them all up." Walsh paused, then rendered the italics visible. "She makes me *look forward to going to work*."

"Even during the difficult parts. The last part of her perfume, when we were splitting molecules, she was learning the industry in a short, intense amount of time, and we were explaining to her that the fragrance she'd been making all those years, the blend she loved so much that she'd been mixing on herself of Bonnie Belle musk, that

Egyptian oil, and the Comme des Garçons *Incense Avignon*, that she'd wanted that so much, and Laurent and Clément had created that for her, and privately Carlos and I knew that her perfume, at least her first perfume, shouldn't be that. So here I was saying, 'OK, you have to make the decision. You have to approve this. Accept this. And we think it should be different. Are you *sure* you want to go there? Are you *sure* you want the scent to be that masculine? We're not so sure. The first is the first. Sometimes you only have one shot.' And she's listening intently, focused. She gets it. As hard as that is. She's learning a new industry, and she has to kill her baby. It was . . . astonishingly tough. We went in a different direction, away from her original dream. And it takes someone with a big mind to process all those things on all those levels—emotional, intellectual, aesthetic.

"The way Sarah Jessica talks about her mom and the way she was raised is very similar to the way I was raised. Very humble background. No mean words. I still call taxi drivers 'sir.' Living in New York and in my job I've developed a much tougher shell than she has. And when I started to spend time with her, she made me remember all the things my mother taught me. It's actually easier now to be a woman than it was twenty years ago. I don't think she has any shell. She says 'gosh.' She asked me where I stay in LA, I say, 'Chateau Marmont,' and she says, 'I'm not cool enough for that.' She's the person who opens the door for everyone else.

"You know, if I'd been acting for three decades and in her position, I think it would be hard to be confused. But she allows herself that. And I think that's really rare. That's not to say she's light, or 'sweet.' We had a 6:50 A.M. flight one time for a personal appearance. She carried her own dress on the plane. We sat together. She was delightful. Got off, she went right into hair and makeup, I sat there watching them getting her ready, and she's focusing on *The New York Times*, absorbing it completely, talked about politics and economics with everyone in the room, then went and spoke to three thousand people who'd come to see her, and we got back on the plane.

"At her personal appearances, people bring notes, flowers, pictures of their children, 'Would you sign it?' *Sex and the City* books. She signs. She writes. She calls people. She called a woman and thanked her. I'd have to drag her away. I'd say, 'OK, we've got twenty-five minutes and a hundred people, you're going to have to jam on it.' She'd say, 'OK, but are you sure we're going to get through *everyone*?'

"When we launched we started to send her sales reports weekly, by account [store]. She'd read them, and if we weren't in the top three, she'd e-mail me back and ask me—and I'll find you one of her e-mails—'Is there something I need to do? Do I need to go there?' " Walsh stopped dead, looked incredulous. "I mean, *come on!* Who *does* this? Even *I* don't do this! But she would! Carlos and I just looked at each other and said, 'She's really going to go to a department store in Dadeland country because we're number nine?'

"When we went on *Oprah,* after we left, I was in a car going to the airport, they'd put her in a different car, she called my cell and said"—Walsh does an extremely creditable impression of Parker's earnestness—" 'Hi, hi! It's me! Do you think it went well?' I said, 'Are you kidding? It was ten times more than we expected.' She said, 'Really?' That reassured her, but then she said, 'I was just so worried when I told the audience that everyone would be going home with a bottle of my perfume that I wouldn't get a response.' "

I write the piece, send it in to Andy Port, my *Times* editor.

A few weeks later, I'm at my desk and Belinda calls me and says, "So, listen, about the exclusive."

Yeah.

"We're still committed to giving an exclusive on the perfume to *The Times,* OK?"

Yeah . . .

"She's going to be on the cover of *Vogue*—"

Uh oh.

THE PERFECT SCENT

"—but"—Belinda moves on smoothly—"it's going to be about her life and her next movies. There'll be, like, a *paragraph* on the perfume. No more."

Uh . . . OK, I say, shouldn't be a problem. (I have no idea, actually. I've never done this and am making it up as I go.)

I call Andy, and she listens, asks a few questions in her concise way, and says fine.

When it hits the newsstands a full month before ours, half the *Vogue* piece, it turns out, is about the perfume, and Andy's reaction is very dark and she uses language I've never heard her use, but that's the business, and in the end *Vogue* has its piece, we have ours. The *Times* piece runs—SJ likes it, Coty likes it, and frankly we quite like it too.

The brands are constantly proposing articles on their launches. Ninety percent of the time they offer no substance and have no intention whatsoever of allowing you to see anything real. So you pass. Around the same time that I'd started talking to Belinda, I'd gotten a call from a PR rep for a major American designer. "New perfume," she'd said.

And? What's the story?

She'd sighed. Her designer is as risk-averse as they come, his communications Stalinist, his marketing regimented, his creative process Kremlinesque. She knew I wouldn't be interested. We left it there.

The night the Parker piece came out, she saw me at an industry event, came over—I like this woman; she's very direct, practical, realistic—and said brightly, "Congratulations, it's gorgeous, and god-*damn* you, that could've been us."

You didn't make the offer.

"Oh, I know, I know," she said, looking around the room. "They're incapable of letting anyone in."

So they don't get the piece.

"So they don't," she said with a tight smile, "get the piece."

⁓

Months later, it's December, and I leave for Paris to do two stories.

All those early mornings of coming-to on airplanes, tight, dry eyes in desiccated air. Raise the shade and peer blearily out through the plastic oval at the freezing coals of sunrise twenty thousand feet over Paris. In New York it's half past midnight, here it is 6:30 A.M., and the sky is beaten black and blue and slit open along a red/amber knife's edge as if it had just been wounded and we are flying slowly by the damage to inspect it. You fly strictly economy on *The Times*. You're sitting up. As you descend into the cloud layer, the lights flicker on painfully and the chief flight attendant's voice announces with iron cheerfulness, "Ladies and gentlemen, the captain has illuminated the seat belt sign indicating we've begun our initial descent to Charles de Gaulle." There are two absolute capitals of perfume in the world, New York and Paris. I live in one, so I visit the other. Before I lived in this one, I lived in the other.

Arriving in Paris means the captain tells you about the latest strikes. The captain announces that last week government services struck, this week it's trains (I hope the car service will be there), next week, he says with a bit of an edge, airport workers. Doing business in France means constantly hoping you'll get the last flight out, as if it were a Third World country with a constantly toppling regime.

The plane lands, shudders in the dark, its belly just above the concrete, and the jet engines jam into reverse. Charles de Gaulle is its usual postmodernist mess, a disaster in concrete and glass that is one of the worst major airports in the world. After passport control I turn my cell on in front of the bored customs agents, and by the time the baggage starts appearing I have a voice mail: "Please meet your driver," says the message, with no indication where.

It's freezing and gray. Sure enough there are no trains. I watch a fifty-year-old Sikh taxi driver in a big blue turban wipe off the window of his Mercedes with a paper towel and then simply throw the towel on the street, just as one does in Mumbai. My cell phone has no more minutes. I look around and ask a friendly Sri Lankan taxi driver if he might lend me his phone. "Certainly!" he says, and I dial the travel agency that arranged for the car service. The woman on the line apologizes: "*C'est totalement aléatoire, les voitures à Paris.*" Car service in Paris is a totally random affair. She tells me to go with the Sri Lankan guy, who is happy about that. I have the gray grit of this Paris winter morning in my veins. I want to brush my teeth. I can't remember where I put my passport.

The taxi is a new Renault, very comfortable, firm seats, smooth synthetic fibers. French cars generally feel great. I sit in front like the Buddha, and he chats at me energetically. I understand little of his English. Alternatively I understand little of his French. There's a cross, a cheap plastic Blessed Virgin Mary, a Krishna image, Hindu prayer beads, a feisty Singhalese good luck demon, and a Tibetan prayer bracelet. All this from the air-conditioning knob.

He attacks my hotel by circling the *périphérique*, the concrete belt around Paris, then dives in via the Porte d'Asnières, negotiates the boulevard Malesherbes. He talks and talks. I understand nothing. Fascinated, I take a scientific approach and start interjecting non-sensical comments at inappropriate moments. He gives no indication of noticing. He drops me at Le Grand Hotel Intercontinental, happily waves good-bye. My bags disappear. My room, which is a seven-minute journey from reception by foot (you need sherpas), has twenty-foot ceilings and looks out on the Opéra. By the time I reach it, I am semicomatose.

My bags reappear in the room.

I go to my appointments. I spend two hours with a lovely young perfumer named Mathilde Laurent. Cartier has just taken Mathilde away from Guerlain, where she created the estimable *Shalimar Light*. Cartier has not yet figured out that, as Ellena and Polge are to Hermès and Chanel, Laurent should be the Cartier in-house perfumer. They have not realized she should create all of Cartier's scents.

Maybe they will someday. I have lunch with Jean-Michel Duriez, the in-house perfumer at Jean Patou, at a nice restaurant. As we're paying the check, I see Véronique Ferval from IFF and stop to talk with her and her husband, the scent sculptor Nobi Shioya, who runs his own house in Brooklyn, S-Perfume, creating astonishing art scents. I leave the restaurant, walk around the corner past the Guerlain store, turn right and go talk perfume for a few hours with Pamela Roberts and Rémi Cléro, the creative director and CEO of l'Artisan Parfumeur. (On trips in summer, I rent a scooter and in T-shirt and jeans bomb around the streets I used to live on. I never go to the museums.) Late afternoon I work out in the hotel gym, eat dinner by myself. The next day I go to Dior, which is starting an interesting project, then have lunch with the perfumer Dominique Ropion at a place on the place de l'Alma, and we talk and talk, perfumes, molecules, launches. That evening I'm invited to friends'. I carry an umbrella around under the gray/black sky whose belly is as swollen with rain as a tick with blood, but it doesn't rain.

Perfumers are deeply strange people simply because their sensorial perception of the world is so highly trained. The educated olfactory capacity makes spending time with them not unlike spending time with talking Labradors. I was on an Air France flight from Florence to Paris sitting with the perfumer Marc-Antoine Corticchiato, and the flight attendant, with the refined, almost prissy movements of many large heterosexual Frenchmen—they look like football players from finishing schools—handed us the *serviettes rafraîchissantes.* I ripped mine open, rubbed the towelette over my hands, stopped to lean into the smell.

Not bad, I said.

Corticchiato glanced over. I held up my hand, and he smelled and made a dismissive face. "Eh."

Wait, seriously, for a towelette in aluminum? It's not bad.

"Linalol," he listed, throwing them off quickly and without interest, "linalyl acetate for the fake bergamot, citronellol and geraniol for the flowery, dihydromyrcenol, Galaxolide, *qui ne coûte rien du tout et qui rassure."* Which costs nothing and which reassures. *"Et c'est ça."* He shrugged. *"Tous ces trucs cheaps."* All these cheap things. He saw the codes behind the images and heard the tones outside the normal human range.

We landed. On the hike from terminal 2D to 2E, we walked by a Charles de Gaulle International toilet, and as we passed through a pocket of invisible molecules in the air, Corticchiato reeled off the raw materials scenting the French toilet-cleaning products.

Perfumers work in the patronage system—the perfume industry is a quintessential patronage paradigm, with Pucci and Kenzo in the role of the Catholic Church and the king of Spain—and they have, since the system began, both embraced and lamented its strangenesses and quirks. The perfumers are emphatically not free. They are employees. They are also emphatically artists. Their strangest relationships are often with each other: competition, collaboration, envy, revenge, admiration. Their lives consist quite literally of creating liquid secrets, doling them out judiciously, hoarding them, playing the politics. They amass molecular treasures at one Big Boy, then another Big Boy makes them an offer (better pay, a transfer to Paris, subsidized housing and tuition for the Lycée français in New York City, whatever it is), and they jump ship. The corporate types loathe it. It's well known that before they leave, the perfumers download their formulae, quietly photocopy charts, scan captives, and send batches of dark via anonymous personal e-mails to hard drives in their home offices. (Captives are patented molecules that cost a premium and are only available from their creator companies, just like nongeneric drugs.)

Then, once you've unpacked in your new office, you get to know your colleagues: other perfumers.

Each perfumer belongs to a corporate camp, and everyone knows what everyone else looks like, and the bosses are always watching, and so at parties, they can but they really can't socialize.

They do and don't talk to each other. It's all sort of vague, and the boundaries are unclear. But they are hyperaware of each other. On his first day at his new job at one of the Big Boys, a friend of mine was grabbed by one of the indigenous perfumers and marched into her office and watched in amazement as she began a systematic analysis of every single perfumer at his previous company based on the way they built their formulae. To his astonishment, she was almost completely right about everything. "X must *love* women! He must sleep with an *awful* lot of them." "Y is manic depressive, right?" "Z is a closeted lesbian. You can smell the frustration." "B has a thing for T; she follows every accord he makes—is she fucking him?" "M has to be completely sexually frustrated; you can smell her closed-up hole in everything she does."

Some perfumers wear an elegant remoteness like armor and take it off only on the phone late in the evening; some are friendly and vapid; some are intense; some hide their ambition. Some are posers. Some are so suspicious and paranoid and repressed that they wear a mental psychiatric suit every day and are incapable of saying anything even remotely interesting. Zombies. Some are wonderful, salt of the earth, they tell you terrific stories about this scent or that one, explain the latest captive, and every minute with them is a delight. Angels. Showmen. Some try for beauty. A few don't.

Francis Kurkdjian is nervous as lightning. Carlos Benaïm is cautious as a Thai diplomat. Bertrand Duchaufour is earnest, Pierre Wargnye is your best drinking buddy, Olivier Cresp and Christine Nagel could each run a French ministry, Daphné Bugey will cut you in half if you get in her way, though she won't mean it. Jean-Michel Duriez is a loving Teddy Bear. Yann Vasnier is less and more than meets the eye. Sophia Grojsman stands behind the Tsar. Dominique Ropion and Harry Frémont could happily fish with you for hours. Jean Guichard melts in your mouth. Anne Flipo is practical, take it or leave it, and Olivier Polge you would introduce to your daughter. Christophe Laudamiel is hypermetamorphic.

Jean-Marc Chaillan will argue with you in a bar. Thierry Wasser has secrets. Caroline Sabas doesn't. Calice Becker is decidedly present. Pierre Negrin, less so. Stephen Nilsen wants things to be right. Dave Apel is game if you are. Ilias Ermenidis is Greek as Euripides, Maurice Roucel is French as de Gaulle, and Alberto Morillas could have sung tango on the radio to adoring millions.

Some perfumers hate their colleagues; some are nervous; some are sweet as chocolate; some have been abused. Some are fascinated by the work; others are burned out. Some are paranoid. Some demonstrate an amazingly earthy, blunt sexuality, a frank cynical crassness that I love.

Freud would have a field day in this industry. Though if this is true, it is because the crucial people in the industry are artists.

On the last night of the Paris trip I go to a party. It is a Cartier party. I put on my tuxedo and go downstairs at 7:01 P.M. I walk down rue Scribe, cross boulevard des Capucines, jog left on rue Daunou for a short block, then right on rue de la Paix, where the Cartier store sits, famously, at number 13.

The rue de la Paix—one of the most expensive streets in the world, with the place Vendôme (begun in the late 1600s) and its baubles, the Ritz, Chanel, Dior, Mauboussin, and the rest on one end and the place de l'Opéra at the other—is utterly transformed. It has snowed, but the snow covers the street between the rue Daunou and the rue des Capucines within a precise rectangle only in front of the Cartier store and the Park Hyatt Vendôme just next to it because it is fake snow. I reach down. It's wet. It squishes out between my fingers like goop, some sort of plastic. Amazingly realistic. It has been carefully laid there by a massive team of workers with shovels and trucks.

Snow falls on our heads, glittering. In flight it looks utterly real. It's snowing in Paris, and it rests in people's hair. They're blowing it from the top of the building, but this isn't the plastic, it's something else, metallic acrylic flakes. A crowd in coats and mufflers and hats is gathered behind the ropes to watch the arrivals, a constant stream of Mercedes and Jaguars dropping off guests in tuxedos and gowns. The fake snowflakes mirror the glitter of the gowns. I wonder if it's cold enough to really snow. There are ten miniature stages that we walk past on the way to the store's entrance, each one a fantastic life-size jewelry box enclosing a costumed person: a Chinese princess, a sorceress, a ballerina. They perform over and over, the curtains going up and down.

The crowd cranes its collective head. There's a rush of excitement—is that Monica Bellucci? Yes it is, her hair up. Searchlights move across the sky, and the camera flashes sparkle against the jewelry dripping from the necks of the women. A very tall, pretty young American woman speaking good French is being led in, the cameras on her. She and I enter at the same time, and she says to her friend, "Oh, meet my agent." The French speak English to the Chinese. Thin, gorgeous, curving young women wear silky gowns covering thighs under fur wraps.

A group of Spanish women sit in the gallery that rings the store on its second floor, women so tall and exquisitely beautiful and immaculately dressed it takes the breath away. One of them sullenly twists a lacquered fingernail into a diamond necklace, untwists it, twists it. People are speaking Mandarin, Russian, Japanese, French, Cantonese, Italian, English, Spanish, German, Swedish. The store, packed with guests, gleams. Aliona Doletskaya, the editor in chief of Russian *Vogue,* is chatting with Suzy Menkes of the *International Herald Tribune.* Someone introduces a countess of something to a CEO of something else. He kisses her gloved hand.

An army of uniformed young men hired by Cartier circulates. "If you would please make your way to dinner," they murmur. We spill out of the Cartier store, turn right down the rue de la Paix, and walk through the place Vendôme, lit and shining. The stream of gowns flows between the Ritz and the Chanel jewelry boutique, runs down the rue Castiglione past the Guerlain and Payot stores, past Jean Patou. Castiglione is entirely colonized by an immaculate flotilla of new Mercedes, their drivers waiting. ("*Mais combien ils ont dépensé pour ça?*" demands a French gentleman to the woman next to him. But how much have they spent on all this?)

The tuxedos flow into the Tuileries, whose forged iron black lacquer gates have (now this actually makes people exclaim out loud, wait, *the Tuileries*? how did they get permission?) been modified, its spikes removed and replaced with a golden word: *Cartier.* The guests pass under the word on the wide red carpet, stilettos and shiny black shoes walking down into the Tuileries and into an immense Cartier-red tent that's been constructed, as wide and long as a cruise ship.

There's a problem at the reception table. The young women in black search fruitlessly for an Arab couple under *M.* They aren't under *M.* No? But your last name begins with *M?* Yes, yes, Meheri, says the nice Arab man, who is wearing a full-length black ostrich skin coat and a mustache, the one in worse taste than the other. Finally the Cartier people find them: They're under *A.* Oh, yes, says the Arab man, Al-Meheri. Everyone smiles, and they sweep inside. Photographers are taking photos everywhere, it's like a military bombardment, the flashes illuminating the amazing breasts mounding out of dresses of velvet, satin, lace, feathers. (They strafe the tall, pretty young American woman. She turns out to be named Electra and is the model for Cartier's most recent campaign shoot.)

Everything inside is Cartier red. It is the deep 1800s French red of Deuxième Empire boudoirs, the red of the velvet in the old *calèches,* the red of a theater seat. Red velvet. Red flowers. Red walls. Red carpet. Red candles. Red drinks. The furniture is trimmed with gold. Two women in black gowns and four pounds of pearls around their necks pose on a sleek red settee like Ingres courtesans. They do nothing, just pose, and everyone gives them space to occupy. A huge screen shows images of Elizabeth Taylor accepting Cartier jewels. Grace Kelly wearing Cartier jewels in the Monte Carlo sun.

The Frenchmen put on their usual show of talking loudly and importantly into their cell phones. When people look at their loud, self-important talking, they glare at them. If people don't look, they find people to glare at. It keeps them busy talking and glaring. Russian women murmur into their cell phones as if putting babies to bed. They smoke and murmur, smoke and murmur. The Japanese search for each other in the vast room on cell phones. "*Ima doko?*" (Where are you?) *Mienai yo!*" (I can't see you!)

First the corps of coat check women, after them waves of shock troops of cocktail girls carrying drinks, the army of waiters invading after that from the left flank. As the Europeans exhale thunderheads, the usual putrid smog starts to coalesce in the room, filtering the cone of light projecting Elizabeth Taylor and her diamonds. Violins play somewhere. Gay men comment on everything in sight. Straight men follow women around. "*François, comment tu vas?!*" The waiters prepare their artillery (three kinds of caviar to start). A phalanx of bellhops in 1930s pillbox hats, dressed all in red, boys freshly scrubbed and beaming, deliver things, scurrying about. Security, head to toe in black, never smile nor even move much, as if the gravity of their job anchored them physically to the planet. The entire rue Castiglione, the entire rue de la Paix, this entire corner of the Tuileries, this one night, this moment seems the center of the world.

Dinner begins around 11:00 P.M.

An immense American swing orchestra, flown in from the United States, occupies a vast stage before the immense dining space. Men negotiate organza trains on their way to the dance floor. The orchestra is excellent. The actor Jean Reno dances elegantly. The daughter of a New York diamond merchant is talking to the daughter of the artist and filmmaker Julian Schnabel; Cartier flew them in first class, put them at the Hôtel Costes. (Everything is on Cartier's tab: minibar, Internet connection, you name it.) A wealthy Chinese woman holds court at her table, reigning in a black velvet rhinestone-covered bolero jacket ugly beyond measure. Two young and beautiful English couples debate the merits of an expensive resort in the Maldives. A tall young Chinese man, muscular and movie-star handsome, walks silently across the dance floor in his tux, followed by the eyes of three Italian women. He is exquisite. The food is delicious.

One of the crisp young Englishmen stops, blinks, looking around. His companions pause. "How much did they spend on this?" he asks them offhandedly.

❧

The next day, I'm at Charles de Gaulle for my flight back to New York. I pass a tall thin African guy whose body stinks so strongly I can smell him twenty seconds after we've passed each other. His stink is complex, nuanced. He perfumes the air like a boat leaving a wake.

At security the big Arab guy patting me down finds my Bigelow Mint Lip gel in my pocket and tosses it, which makes me furious, but the small Arab guy who checks bags misses my toothpaste in my carry-on. As he's missing it I think, I should have put the god-damn mint gel stuff in my carry-on. Then I realize something. I ask him, Are you wearing Azzaro?

He looks at me warily. The two guys on the right and left sides of him look at me. A fat passenger with a cheesy mustache looks at me.

I'm a perfume journalist, I explain.

He's like: *Huh.* Weird. OK. "*C'est Azzaro,*" he says. He says it a little proudly.

Is it *Chrome?*

He frowns, thinking. "I don't know." He says, "My wife gave it to me."

"It's not *Chrome,*" says his security colleague to the right of him, rubber-gloved hands forearm-deep in the mustache's bag. He says it almost scornfully. "I'm wearing *Chrome.* His is something else."

"Do you like it?" says my guy, to me. (He ignores his colleague.) He's looking hopeful now. "You're an expert in perfume?"

I write on it.

"Vous aimez?" Do you like it? He means his scent.

He is a sweet guy, and he's wearing this scent that his wife gave him, and I can smell it on the other side of the security table with my lip stuff in a trash bag below, and it's a vile smell, as if there were a chemical fire in Terminal 2E and the steel and plastic were a bubbling stink around us. I think about all the great scents I could tell him about, why they're great. I think, quickly and without interest: linalol, dihydromyrcenol, Galaxolide, *qui ne coûte rien du tout.* All these cheap things. He's hopeful. I say, It's a classic masculine. He shoots a knowing look at the guy on his right. He's got a classic masculine. The mustache nods sagely: Azzaro.

We say good-bye, he waves, and I head to gate E77.

My cell phone rings. It's Belinda Arnold from New York.

"So guess what," she says cheerfully. "You up for another one?"

What've you got?

She asks me how would I like to sit in on the creation of Sarah Jessica's next scent product.

The next perfume?

She hesitates. "Sort of. Maybe. A perfume product we're going to be creating that will use *Lovely.*"

I'm not sure I get it.

She says, "Well, we're thinking about it now. But whatever it turns out to be? We'd like you to come along for the ride. We'll include you in the whole creation process."

Parker and the perfumers?

"Yep. But following the story not from the perfumer's point of view." She knows that in my reporting I head reflexively for the perfumer's point of view. She wants to start us off clear. "It'd be from her point of view. Inside what it's like to go through the creation of a perfume as the creative director. What do you think?"

I'm in.

CHAPTER 5

E LLENA LEFT PARIS and flew home to Grasse, where he
took a call from Gautier. She told him that they had
definitely accepted AG2 as the general theme. "*Ça sent
bon*," she stated. It smelled good. "*C'est ce tableau là qu'on veut.*" This is
the picture we want.

And now, she said efficiently, changes. She ran down the list,
which he found at once a lot and not much. They wanted him to
keep the freshness and the *fusant* aspect (the word is related to *un
fusil*, a rocket), the sparkling quality, but Dubrule found it a bit too
tranchant (to her it was a love/hate scent). Gautier thought it was a
bit too harsh grapefruit. They both felt there was a problem with
the *persistance*, which is the amount of time the fragrance lasts on
skin and a common technical problem perfumers deal with; the
juice was not as tenacious as it should be. And they wanted the *fond*,
the base, to be more present.

He hung up encouraged, basically happy. "I have no anguish once I've got to the '*ça sent bon.*' " He paused. "Well, for a few minutes I have no anguish."

Ellena, by his own account, works within two alternating psychological systems, extreme doubt and absolute certainty. His wife, Susannah, knows both quite well. "I have," he says, "terrible, terrible periods of doubt and fear where I feel like hiding under the table, and other periods where I say, '*There*—that is exactly where we need to go.' "

At the meeting in Paris he hadn't admitted it to them, but they'd surprised him. He'd thought they'd choose AG3, the magnolia-inflected green mango, for which he had a certain *faible*, a weakness, and not AG2, the lotus. In fact when he'd said to Gautier, "I know which one you don't like," he'd thought it would be AG2. When he realized he was wrong, he didn't say anything. "I know her well," he said of Gautier, "but I don't know everything about her." He made a mental note.

He'd thought they'd choose AG3 "because it was the strongest character." Sitting in his lab, he made a little moue, looked pensive for an instant, then dismissed it. "*Bon, c'est pas bien grave.*" Well, it wasn't a big deal. AG3 was, he admitted, quite pronounced. "It's difficult to accept that AG3 is so pronounced," he said. "But . . . still . . ."

If you are the creative—if you are Sarah Jessica Parker and have a scent in your mind that you want the perfumer to find for you— and you fail to communicate your vision, you're building a disaster. If you are the perfumer, every clue the creative gives you, every casual comment they make about every single smell, is to be ignored at your peril. Ellena reassembled the Egyptian comments in his mind.

When they'd been in the Nile village, just discovering the mango trees, he remembered Dubrule pressing her nose into the branches, inhaling the marvel of natural olfactory technology. She had remarked, surprised, on the carrot angle. "A fruit that smells of a vegetable?" she'd asked. Yes, Ellena had replied, what you're smelling are molecules grown in both carrots and mangoes.

Now in the lab he said, "Naturally I put in carrot to get my mango. But that's a special trick of mine. There are very few perfumers who can do that. Other perfumers say, '*Merde, il a trouvé cette astuce-là; je n'aurais jamais pensé à ça!*'" Shit, he found that trick; I'd never have thought of that! Dubrule had also gotten a hint of apricot and grapefruit, and Ellena knew why. "I can tell you exactly which molecules that thing contains," he said. "I can analyze things faster, and I can put things together faster."

In Egypt he had watched Gautier frown under the green mangoes, and . . . they smelled like . . . *nail polish remover*? Yes, Ellena said, green mango contains the molecule acetone, aka nail polish remover. "You will above all not put nail polish remover in the perfume!" Dubrule had said to him. "Above all!" Gautier stated, looking at him. Of course I won't, Ellena had said to them in Egypt. He knew full well that he would, "but," he explained now in his lab, "in such a manner that you don't feel it." Perfumers often put acetone into their formulae to give a sort of shimmering/lightening effect; there were things you did that you didn't necessarily tell the creatives about. "I put incense in *Nil* to get the mango resin smell, but what actually interests me is not the incense itself but what it does to the rest of the perfume: If you don't put it in, the perfume is very cold." The politics of the process bothered him not at all. He just had to be a little diplomatic. "There's always a *décodage*," a decoding, "between what they say, the scent in their minds, and what I am actually constructing."

Philosophically, Ellena is a devout minimalist in terms of materials and a maximalist, in the classic French intellectual manner, in terms of thought. He practices (currently, not originally; he has evolved into it) a sort of Bauhaus school of perfumery: clean lines, deceptively neat structures, simple formulae, luminescence, clarity. His style took the entire history of Western perfumery and distilled it to its highly thought-through essentials, then carefully mutated each one of those forward until it reached a modernized format. One could say the approach simplified without necessarily purifying.

Ellena thought about things like: What fragrances does the human species react to and why. He said things like: "*Ce n'est pas l'argent qui m'intéresse. Je suis un homme d'idées.*" Money doesn't interest me. I'm a man of ideas. "*Je défends les idées et les principes.*" I defend ideas and principles. "*Ça pourrait apparaître stupide à certaines gens.*" That might appear stupid to some people. He shrugged. He'd begun, a young, freshly minted perfumer in the Grasse office of Symrise, by making—his very first scent as a professional perfumer—*Eau de Campagne* for Sisley, which launched it in 1974. It was one of many perfumes that year, and, as usually happens, no one noticed. (It can take five, ten perfumes before having even a half-decent success. Or never.) Ellena's second perfume he made for the jewelers Van Cleef & Arpels. He did it in the ultraclassic, traditionalist French style with all the heavy gilt rococo flourishes, the olfactory equivalent of plush scarlet brocade. They named it *First* and launched it in 1976. It was a huge hit. This can be disconcerting, a success so young.

In 1982 he created the artistic perfume *La Haie Fleurie du Hameau* for the niche Parisian house l'Artisan Parfumeur, which he did, by contrast, in a rough sketch impressionist style like a painting of a Provence farm by Monet: delicate, sentimental, a bit wistful, a bit comfortable. All in all, by this time he'd won around twenty briefs. In 1984 he did Bulgari's *Pour Femme* (the original version; it was changed later with a relaunch); in 1990 for Rochas he'd gone back to the French traditionalist school for *Globe;* in 1995 he'd changed stylistic gears and done a commercial scent, *L'Eau by Laura Ashley;* then for Yves Saint Laurent he switch-hit with the gorgeous commercial electrical neon *In Love Again* (1998), which was like smelling a three-story-tall gas plasma screen over the Champs Elysées showing a bowl of the ripest tropical fruit. In 2003, collaborating with the Symrise perfumer Lucas Sieuzac, Ellena had produced a typical Armani commercial pastiche for Emporio Armani, *Night for Her.*

His biggest hit was his most unlikely. In fact the story of El-lena's *Eau de Bulgari* is one of the stranger episodes in perfumery.

When he'd done *First,* this classic rich French perfume, he'd packed in some 160 materials. But by the mid-1980s he felt strongly that it was time, as he put it, "for me to show that *I* have something to say in perfumery, not just what you ask me for." Jean-Claude and Susannah are serious lovers of tea and began frequenting the tea seller Mariage Frères, which at that point hadn't become as famous as it is now. He went often and loved the smells—walking into a Mariage Frères store is categorically one of the most exhilarating olfactory experiences it is possible to have—and after making his purchases he would go back to his lab and, all but compelled, write short formulae sketching the scents in his head. He kept the sketches in a drawer.

One day he went and asked them if he could smell their teas—Moon Palace, Lung Ching Impérial, Earl Grey, French Blue. They agreed, and he spent a whole morning smelling all hundred of the large metal canisters. He was developing an idea, refining it to a point. His *astuce,* his trick, was to mate a synthetic called ionone, which had as far as Ellena knew been used only to make the scent of violets, to a molecule called Hedione, a synthetic derivative of methyl jasmonate, the natural—and key—odorant that you smell when you smell jasmine. Hedione is a gorgeous molecule, a man-made beauty with an ethereally unidentifiable scent, like olfactory halogen light in a liquid form. From these two molecules he made a scent of tea, though not a particular kind of tea; it was, as Ellena would be careful to explain to you, actually the *concept* of tea.

Dior was then asking for submissions on a major new Dior masculine to be called *Fahrenheit,* and Ellena submitted his tea fragrance. Dior loved it and, after changes and questions and a lot of waiting, told Ellena that he had won the *Fahrenheit* brief. He went to Paris and drank champagne with the Dior bosses, and everyone celebrated. The next day, Dior called to say they'd changed their minds. The marketers were uncomfortable with the abstract, un-known quality of Ellena's juice. They announced that the winners of the brief—one of the most important of the decade—were the perfumers Jean-Louis Sieuzac and Michel Almairac, and their sub-mission became *Fahrenheit*, which turned into a huge hit. Ellena was astounded, then crushed.

He tried to rally. He took his scent elsewhere, and word got back to him of the mocking comments of his colleagues. "Have you heard about Ellena, taking his little tea scent around Paris? Yes—*tea*." According to Ellena, the decision makers at Yves Saint Laurent passed on the juice, saying, No, it's not for us, it's too creative. He tried other houses, urging them, "I think it's really something new, it will work." But none of them wanted his scent.

At that moment, knowing nothing of this, some executives from the Italian jeweler Bulgari approached him. They explained that they had been envisioning a nice fragrance that, perhaps, would be sold in some quiet corner of their store, an eau de cologne maybe. Oh, and make it 100 percent natural. The scent's primary role would be to perfume the boutiques, give them a pleasant smell, though yes, certainly, it would also be worn by the few clients who might buy some now and then. They did not at all think of the fra-grance as a product that would bring in money; this was simply about another part of the identity of Bulgari. Ellena took a deep breath and, first, said that yes, well, an eau de cologne might be nice, but you don't want a 100 percent natural because all that will give you will be a standard eau de cologne exactly as it was in the eigh-teenth century, no lasting power on skin, no originality, the same as a thousand others. Second, I have an idea, and he brought them his draft of the tea scent.

Eau Parfumée au Thé Vert de Bulgari, which became the name of the perfume he ultimately did for Bulgari, is a smell as deep and strong and clear as Turkish seawater. The scent has power, a technical feat. Aesthetically it conjures a small amount of the smoothness of

Darjeeling but gives in much greater proportion a rough, potent black tea from China; Bulgari's marketers called it a green tea, but it has only the freshness of green tea, not in any way the scent. There is a vaporous trace of old wood smoke from the fire used to boil this pure water, and at the same time the scent is shot through with this freshness, which is why, as Ellena intended, it smells like tea and, simultaneously, it doesn't. His idea was—explicitly—not to copy reality. His idea was to transform reality. Shortly after the scent launched, the manager at Bulgari Parfums called Ellena to report, in a strange tone of voice, that, odd, but at the Bulgari store in New York they were selling ten bottles a day. At $350 per bottle. Once the Bulgari executives realized what they had, they sent it into the market, and it made buckets of money. This was a perfume never meant to be distributed.

"When I did *Eau de Bulgari*," Ellena told me, "everyone said, 'Oh, another cologne.' And in two years everyone had copied it—it gave birth directly to *cK One*. They didn't give me a brief by the way. They'd told me they wanted something *discret, léger, et qualité*" discreet, light, and of quality "and what I brought them was completely mine."

"I have two periods in my life," Ellena said, "from *First* to *Eau de Bulgari*, from 1976 to 1992, then after. I was young when I created *First*, and I put everything others had taught me into it. With *Eau* I had my own story."

At the time Hermès began courting him, Ellena was just completing a perfume called *L'Eau d'Hiver*, the third (and, given his entry into Hermès, obligatorily the last) of a trilogy of perfumes he'd made for an exquisite, in fact all-but-revolutionary niche collection, *Editions de Parfums*, run by the scent impresario Frédéric Malle. Ellena's approach to the construction of *L'Eau d'Hiver* reveals the technical and intellectual approach he had evolved by this point in his career.

He began his perfume by intellectually reconceptualizing the great Guerlain classic *Après l'Ondée*. "The problem," he started and then immediately checked himself. "*Well*, you can't say there's a *problem* with *Après l'Ondée*, but . . . *bon, voilà:* They were too opulent, these Guerlains in the baroque, supercharged tradition of voluptuousness. '*J'en mets, j'en mets, j'en mets.*'" I put in, I put in, I put in. "At the same time, there's the Guerlain *sillage.*" *Sillage* is the scent wake the perfume's wearer leaves behind in a space, an olfactory infrared arc of their trajectory. It's a technical aspect of a scent that the perfumer must skillfully engineer. I believe it was the perfumer Calice Becker who once defined *sillage* to me, somewhat metaphysically, as the sense of the person being present in the room after he or she has left. "It's marvelous," said Ellena, "this gauzy veil that envelops you. So I wanted to find this *sillage*, but in enlightened form. My style is perfumes that are at once light and very present.

"So for *L'Eau d'Hiver* I took my inspiration from *Après l'Ondée's* theme. Cloud. Soft, comfortable, light, and very present, but without all this"—he gestured—"*grosse étoffe* that you have with the Guerlains, all this *stuff*. It took me forever to do it. The idea of the diaphanous." He conceptualized sleeping in hay in the summer. Heat. Sun. A powder that envelops without weight. He began the perfume's construction with a gorgeous absolute of hay, one of the most sublime of all perfume materials. Hay is, as literally as possible, the smell of liquid summer sunlight. He wanted to create with it the scent of a cloud filled with sun. People expected *L'Eau d'Hiver* to be a cold water (the name means "winter water"). In fact, he was building the opposite, a hot water for a cold winter. He then took an old synthetic, Aubepine (an anisic aldehyde), which smells like a mix of the finger paint you used at age five and the cleaning wax applied to formica floors. Aubepine costs almost nothing, around three euros per kilo. He bolted those two to methyl ionone (a synthetic that gives the idea of iris), the milky-musky molecule MC-5, and a natural absolute of honey. It took him two years to fine-tune this engine, Malle giving him creative feedback, the two of them going back and forth, and in the end, his formula, according to Ellena, totaled twenty ingredients, relatively minimalist for a perfume. *L'Eau d'Hiver* smells of ultrafine ground white pepper and extremely fresh, cold crab taken that instant from the ocean. It is a brilliant, marvelous, utterly strange perfume, unique—it references nothing—and among the greatest ever created.

Véronique Gautier appreciated this maximalist minimalism. As a creative director (the role she played at Hermès), Gautier was ahead of the game in several ways. She understood, as some did not, that one could stuff a perfume like a sausage with the most expensive ingredients in the world and wind up with carefully macerated shit that cost a fortune. What mattered was the way the thing was built.

Gautier was after great art, and great art is great artifice and great manipulation that forces its experience on the viewer, tells him its story, and thus changes him. Ellena was an artist. That's why she wanted him. And Gautier would also tell you instantly that Ellena wasn't there to re-create nature. As an artist, he was principally an illusionist, a description he agreed with emphatically. "Picasso," Ellena liked to say, "said, 'Art is a lie that tells the truth.' That's perfume for me. I lie. I create an illusion that is actually stronger than reality. Some people are surprised when I say this, but: sketch a tree, and it's completely false, yet everyone understands it, and actually if you do a very, very basic sketch, abstractly, people will understand it *better* than if you do every single leaf. Give them too much, and people will start to say, 'OK, it's a tree, but is it an oak or a maple?' " He thus practices a perfumery of very few ingredients. And, he would tell you, the greater a perfumer's power over his art, the greater his mastery of such illusion.

Ellena will dip a *touche* into a molecule called isobutyl phenylacetate, which smells sweetish/chamomile blossom and vaguely chemical, and another into a synthetic molecule whose common chemical name is ethyl vanillin. (Its IUPAC name is 3-ethoxy-4-hydroxybenzaldehyde, and this rich gourmandy vanilla molecule is the heart of *Shalimar*.) He puts the *touches* together and hands them to you. Chocolate appears in the air. "My métier is to find short-cuts to express as strongly as possible a smell. For chocolate, nature uses eight hundred molecules, minimum. I use two." He hands you four *touches*, ethyl vanillin plus natural essences of cinnamon, orange, and lime—each of these has the full olfactory range of the original material—and you smell an utterly realistic Coca-Cola. "With me," says Ellena, "one plus one equals three. When I add two things, you get much more than two things."

He will hand you a *touche* that he has sprayed with a molecule called nonenol cis-6, which by itself smells of honeydew melon or fresh water from a stream. He'll then hand you a second *touche* with a natural lemon on it, direct you to hold them together now, and suddenly before you appears an olfactory hologram of an absolutely mesmerizing lemon sorbet.

The explicit point was not to create a thing but an illusion of that thing, an olfactory alchemy. The point of *Nil* was not to create a green mango but the illusion of a green mango.

He was in his lab. It was June 11. Ellena began the work of changing AG2.

As do virtually all perfumers, Ellena works in his office at his desk, running molecules through his brain as he stares out the window, jotting down the ideas, ten milliliters of this, twenty of that. He sends these down the hall to his lab technician, who assembles them, brings them back, and sets them before him. He smells this, the assemblage he imagined, and frowns irritably or laughs with delight and surprise or narrows his eyes in fury or frustration. And then he adds X milliliters of other materials to another formula on his computer screen and with the push of a button sends those off to the lab.

Like young French chefs dutifully imbibing the culinary canon—with the basic *mise en place* (flour + butter + cream + stock) you master the basic white sauce—all young perfumers can recite in their sleep the recipes for the classic categories. Ellena had learned them at a tender age.

How do you make a basic chypre? ("Chypre" is one of the classic perfume categories.) Answer: patchouli plus labdanum (a species of shrub; its scent is bizarrely animalic, like the fur coat of an unwashed muskrat) plus *mousse de chêne* (oak moss); bergamot as well, if you want.

How do you make an amber? Labdanum + vanillin + Ambroxide.

Junior perfumers discover that Vetiver Huile Essentielle from Haiti smells like a Third World dirt floor and Vetiver Bourbon from Isle de la Réunion smells like a Third World dirt floor with cigar butts. (They hope to do something wonderful with the cigar butts.) They learn, as Ellena knew from decades of work, how to create the illusion of the scent of freesia with two simple molecules, both synthetics: β-ionone + linalol. And orange blossom: linalool + methyl anthranilate, which by itself smells like blossom + aspirin. The classic Guerlain perfumes often used a resinoid material called styrex, which smells of olive oil pooled on a table in a chemical factory. Add 2-phenylethyl alcohol, one of the main molecules in rose, and you get lilac. Add the smell of corpse (indole), you get a much richer lilac. And you can give your lilac, freesia, and orange blossom a variety of metallic edges: Add allyl amyl glycolate, you get a cold metal freesia. Add amyl salicylate, and you get a freesia with the smell of a metal kitchen sink dusted with Ajax powder. Lauric aldehyde C-12 adds an iron with a bit of starch still on it.

A small but increasing number of perfumery raw materials are controlled substances—you see "USDEA" (United States Drug Enforcement Administration) and a warning on the label—because they are the precursors for drugs such as methamphetamine. Security measures are taken, in particular at the larger companies because by definition they have larger amounts of the stuff sitting on shelves and, as one lab tech put it to me, perfume labs are potential crystal meth labs.

By Ellena's rough estimation, the changes Gautier and Dubrule had asked for involved inflecting between 5 and 25 percent of the formula. The estimate hardly mattered; it would depend on a thousand things, and anyway he wouldn't know for sure till they'd reached the final perfume. Ellena assembled the extant AG2 on a fresh standard formula sheet, briskly listing each material, its product code, its precise amount in milliliters and its price per thousand milliliters. Together the materials totaled 100 percent of the formula.

He spent some time eliminating the *tranchant* bitter acidic ingredients, then corrected the proportions. Then he sat at his desk and stared at the formula.

He started with Hedione, which has a jasmine scent. Hedione (its more formal name is methyl dihydrojasmonate) is a molecule that was found in 1961 by searching molecularly through jasmine, and it is a material Ellena loves. He then put in methyl anthranilate extra 10 percent DPG, Iso E Super (a synthetic with a light woody scent used prominently in *Beautiful* by Estée Lauder and Calvin Klein's *Eternity*), and a natural essence of neroli. As a solvent, ethanol, the alcohol in vodka and gin, is used for perfumes, but here Ellena used dipropylene glycol because, due to archaic European regulations, the *concentrées* of perfumes can't be transported from place to place if they have alcohol in them.

He tried three different iterations, smelled them, and didn't like them at all, so he tried four new directions, which Monique, his lab tech, mixed and lined up neatly on his desk in tiny vials. He labeled the *touches* with a pencil, AJ1, AJ2, AJ3, AJ4, dipped each, and held the four spread like a hand of cards in poker.

He leaned over in his chair, closed his eyes, and smelled each systematically. He grimaced thoughtfully. He bent each of the *touches,* resmelled each systematically, precisely two seconds per *touche,* then reversed back down the line. Murmured, "*C'est pas vrai...*"

AJ2 was the freshest, AJ3 the most mango, sweet. He lowered

the grapefruit synthetic in all of them and then added to all except AJ3, in different proportions, *trans*-2-hexenal, a synthetic that smells half of golden apple, half elementary school glue paste, because he wanted a greener fruit. He inflected AJ2 with 5 out of 10,000 parts of Ambroxide, an extremely expensive synthetic derived from clary sage that had been molecularly futzed with. The Ambroxide was for Gautier's *fond*, in theory, although personally Ellena thinks the olfactory pyramid, the cliché glossy diagrams of top, middle, and bottom notes salespeople mechanically deploy at Macy's perfume counters, is "complete bullshit. I'm sorry, you add something to the bottom and you influence the top notes, and when you first smell a perfume you smell everything, top to bottom, instantly."

(As an intellectual exercise in what one must *not* do, he added the peachy *gamma*-undecalactone to get AJ5. As he suspected, AJ5 was now a rich peach that in its lusciousness completely drowned the green freshness.)

His cell phone exploded on the desk, and he jumped, grabbed it, and glared at the number. Brightened instantly. "*Ah, c'est Gautier!*" Answered with a grin and an insouciant solicitude: "*Oui, madame la présidente.*" They chatted. He clicked off, put the phone down, and stared out the window.

Monique brought the new iterations, and he dipped and smelled. Smelled them again. Tossed the *touches* on his desk and sat back. "Ambroxide helps the tenacity," he murmured to the window and scowled. He picked up one of the *touches* again and smelled it and moved his torso back and forth very, very slowly, forward, back. Tossed the *touche* on the desk and looked at the ceiling and smiled a small dark smile. Unconvinced, perhaps. Or perhaps waiting for a material to fall from the sky. Monique brought an envelope. An invitation from a very expensive luxury goods maker. Black tie. He looked at it, chanted out loud, "Monsieur this-and-that, Madame this-and-that would be ravished to have the company of Monsieur

Ellena . . ." He tossed it aside. "I rarely go to those things. I'm not a sophisticate." (*Un homme mondain.*) "I go, I feel like a penguin."

He smelled his iterations again. "At the moment," he said, "I like AJ3. It's the freshest."

~⊗~

That afternoon, Ellena got in his car and drove over to Laboratoires Monique Rémy, a small company in Grasse that is perhaps the most rarified supplier of natural perfume ingredients in the world. Ellena had heard that LMR had a new quality of narcissus called *absolue narcisse de distillation moléculaire*—molecular distillation narcissus absolute. To build *Nil*, Ellena was wondering if a very raw green scent might work well. He was here to smell it.

LMR's tuberose is legendary; its price, like many LMR products, is breathtaking, though the company sells quite reasonably priced products as well. Price is simply a function of scarcity of the raw material multiplied by the difficulty of making the essence or absolute. LMR supplies Chanel with materials it reserves exclusively for Chanel, and no other house can get its hands on them at any price; one of these is Iris Naturelle 15 percent 4095C, which costs $8,500 per kilogram.

Ellena drove up the LMR factory's steep parking lot and parked his new Citroën precisely. On returning from Paris the previous day, he'd picked the car up and found its battery dead. He'd left the headlights on. "This car can do everything," he said, "wash your hair, make you a cappuccino, except turn off its goddamn headlights when the battery starts to die. The one practical thing it might be capable of." He looked skyward. "It's a French car." He walked to the door in the bright sunshine, pressed the button, and stood back, looking up at the factory. He looked intensely interested. "This is the first time I come here as Hermès," he said. He squinted at the building. "We'll see how they treat me."

Frédérique Rémy, Monique's daughter and then commercial director of the company, greeted him. Rémy was an attractive, direct woman with dark hair, a palpable intelligence, and a quick, interested manner. "*Félicitations!*" she said—congratulations—and they both grinned. She and Ellena are both *grassois*. She all but grew up with him, and she had already made him several wonderful things with particular properties he'd asked for.

Several years earlier Ellena had asked Monique Rémy to develop a special bigarade, a special grade of bitter orange, for him to use in a perfume he wanted to make for Frédéric Malle. Ellena loves bitter smells, but bigarade oil has two problems—it is naturally highly red, an unstable color that deteriorates over time, and it renders human skin light-sensitive—which have kept it from being used in large quantities. Rémy made him a colorless bigarade that does not photosensitize, and with it he created a Malle fragrance absolutely stuffed with the material and called, simply, *Bigarade*, an astonishing work of sorcery, an elixir of bitter orange, citrus, white smoke, and the scent of the stratosphere.

LMR had for years supplied Hermès with several of the ingredients for *24 Faubourg:* absolute of orange blossom, an LMR jasmine sambac, an LMR orris butter, and a rose called *rose de mai* (it blooms in the south of France in May). Both Ellena and Rémy were perfectly aware that he would now directly influence LMR's fortunes as he would decide whose vetivers and jasmines were to be specified in Hermès's perfumes. He had choices. Robertet S.A., just up the hill, made excellent things, as did Mane nearby, and the Biolande products had premier reputations. To the degree to which his perfumes were successful, they would demand more materials. Supplying a top-ten bestselling fragrance could mean selling tons of Rémy's Basilic Grand Vert Absolue at €790 per kilo.

To its credit, LMR is spoken of in the industry in the slightly cloaked tones with which Oppenheimers discuss their diamond mines. LMR's factory in the center of France produced, from its own proprietary fields of flowers, a narcissus *absolu*. Only LMR had it. It retailed at around €6,500 per kilo. In Vietnam Rémy had contracted with a French botanist to find unknown plants to develop into new perfume materials. One of the first products that resulted from LMR's new Vietnamese fields was a rare basil with a strange, complex, fresh angle. An Estée Lauder might, for a new Michael Kors (Lauder owns Kors's license), buy up 100 percent of the supply of an LMR material, simply take it off the world market, in order to make Kors's perfume unique. The basil cost €250 per kilo.

LMR is also coveted for its quality control. The materials producers are notorious for trying to pass off cheap Indian or Brazilian distillations as Grasse flower absolutes. "I have to be very careful," Sylvie Chantecaille, head of her eponymous niche house, told me. "There is one house"—she named it—"that is not always good. Their Grasse jasmine is eighteen thousand dollars per kilo, the most expensive in the world, and they recently sent me a complete joke. It was *clearly* from Morocco." She rolled her eyes. "This happens all the time. I find the two most honest companies are Biolande and LMR. Monique has some amazing stuff. Just incredible. Out of this world. There's another company"—she named that one—"that has a rose water concentrate that's two thousand dollars per pound, but don't identify it; I want to keep this secret because I use a lot of it, and I don't want people to know where I'm getting it. In all Chantecaille skin creams, we use rose water, and you have to heat the product, ah, but if you heat it you kill the rose, so you have to . . . stop taking notes and I'll tell you."

Rémy called up to Bernard Toulemonde, the company's CEO. Toulemonde brought down several pairs of industrial protective glasses, and the three chatted animatedly as they did a casual tour of the facilities. They walked past impressive metal-and-glass machines that looked like dinosaurs with tubes and cables sprouting from them. Rémy hated the oversized glasses, which were required by both company policy and French law. She kept taking them off, and Toulemonde kept patiently giving her his lecture on the corrosive chemicals, the solvents, the high pressures, and the extreme heat sources from the machines that surrounded them—which she knew perfectly well. "We can't ask the employees to wear them if we don't," he said. She glared at the ugly plexiglass and demanded of Ellena, "Jean-Claude, can't you get Hermès Eyewear to do something in these?" He promised he'd have Jean Paul Gaultier get on it.

Rémy and Ellena walked around the huge machines, many twenty feet tall. Some had giant metal blades to hash grains and roots. LMR's machines make two classes of perfumery raw materials, and the classes are determined quite simply by the way in which the materials (sage, ylang-ylang flowers, tree moss) are treated: steam or solvent. *Essences* are what one calls scent material extracted using steam at 120 degrees Celsius (almost 300 degrees Fahrenheit). This is "hot" extraction. *Absolus* are materials extracted with solvent (the most common is hexane, which is a little like dry cleaning fluid) at 30 degrees Celsius (90 degrees Fahrenheit). "Cold" extraction. (Actually, solvents will first extract from the ylang-ylang a greasy, gorgeously stinking waxy goo Rémy and Toulemonde would call a "concrete." The concrete is given a *lavage à l'alcool*, an alcohol wash, to dissolve it, and *that* is a liquid absolute.)

If you hot-distill roses with steam to get an essence, or cold-extract them using solvent to get an absolute (perfumers are extremely picky about matching these verbs and nouns), even if it's the same rose from the same field, your rose essence and rose absolute will smell quite different. This is because the two methods take different groups of molecules out of the flower. Distill, say, a species of rose called *Centifolia* (it's one of the only two rose species used in perfumes; *Centifolia* means "one hundred petals") with hot extraction, and you'll get the rose material's lightest molecules, the top notes. Wash another batch of the same *Centifolia* in solvents, and you'll get all soluble molecules including (notably for your smell) the heaviest molecules that give you bottom notes. (Solvents

also remove colorants.) And a perfumer has to be able to manipulate the two. Ellena began his career among these machines as a
teenage *marmiteur* (slang; a *marmite* is a large metal pot), extracting
jasmine, and he can, he says, tell you the country of origin of a jasmine essence. He can also tell you what kind of machine distilled
it: inox (stainless steel), aluminum, or steel.

LMR extraction is widely regarded as superior, and its prices
reflect this. A sample of the Laboratoires Monique Rémy price list
reads, in prices per kilo, like this:

Basilic Essence	€47
Styrax Resinoide	56
Vetiver Java Essence	100
Mousse Arbre Absolue DM 85% IPM IFRA	129
Vetiver Haiti Essence	142
Basilic Grand Vert Absolue	790
Camomille Romaine Essence	857
Rose Bulgare Absolue	1,639
Cassie Absolue Egypte	2,093
Tubereuse Absolue Inde	4,546
Ambrette Absolue	6,792
Rose Turque Essence	6,868
Iris Naturelle 15% 4095C	7,340
Rose de Mai Absolue	8,381

Ellena was not unconscious of price, but he was quite insistent
that "you must never forget price is not axiomatically quality. That
means..." He hesitated, looked a bit foxlike and a bit wary. He
glanced sideways. "It's delicate to talk about such things. How would
nonexperts understand this?" He thought about it. "Take a jasmine
from India," he proposed. "That comes in at around €1,510 per kilo.
These are very expensive products, and there is Indian jasmine in several of the Hermès perfumes. Then take a jasmine from Grasse.

That's around €30,000. When you smell them next to each other, the Grasse jasmine is a bit better. Twice as good. But not twenty times. At that price, you've simply lost touch with reality. It's extremely bothersome for the people who work here"—he looked back at the factory that produces such stratospheric materials—"but at a certain moment, economic reality..." He shrugged. "Look, personally I instantly recognize the difference, but virtually no one who wears the perfume ever will."

He frowned. "It's very important to understand that the price of perfumes is not the price of their materials. You pay for the creativity." He lifted his shoulders, made a rather Greek gesture: He meant himself, the guy right in front of you. "You don't pay for a painting the value of the paint the guy's thrown on the canvas. You pay the years of experience. *Tu me paies, quoi!*" You pay me! "That has value as well, a value that I put into the perfume. This Marxist idea that the price of a thing is the price of its materials is *false.*"

The last room, lined with refrigerators, was the stockroom. Rémy brought out the *absolue narcisse distillation moléculaire* and set it before him. He leaned over and inhaled. She watched him carefully. It was a beautiful scent, an ethereal yet muscular flower with a raw green angle, and it was, he realized, wrong for *Nil.* Too raw, not sufficiently tender. He thought, Shit, it doesn't work.

He got in his car, went back to his lab, and thought, Now what am I going to do?

~～©

The degree to which the perfume industry runs on lies, strange outmoded and anachronistic business practices, vapid publicity, gross hypocrisy, and a general miscomprehension by marketers of what perfume actually is is jaw-dropping. But it is not entirely illogical. Ellena is as fluent in the industry's strange black holes and bizarre features as he is ambivalent about discussing them on the record. The central problem is the formula. The formula is a beautiful thing for those who understand it and terrifying to those who don't.

Creating perfume is exceedingly complicated. It is an art form that is, for example, infinitely more complex than, say, making clothing. Cutting silk crepe into a dress means a piece of silk crepe cut and stitched—expertly, we can stipulate—into the form of a dress. Add the neurosis necessary to get people to spend three thousand dollars for the thing, and you're done. Perfume, by contrast, is, fundamentally, mastering organic chemistry, and it involves cutting and sewing together pieces of the periodic table of the elements, trying to choreograph electrons that often react to each other in surprising ways, and cajoling molecules into a single mesh that has structure, durability, and stability—not to mention beauty, originality, and commercial appeal. It takes all of this to create a formula.

But the thing about formulae that creates real *panic* in the perfume industry—and this is where Ellena's ambivalence in talking to the public is noticeable—is the synthetic molecules. The industry as a general rule is blindly and adamantly convinced that the public will only buy perfumes it believes to be "natural." Since on average perfumes contain 80 percent synthetics, the industry lives every day terrified that the client won't like reality, which thus needs to be suppressed at all cost.

I was at breakfast in Paris at one of the stupidly expensive Alain Ducasse places with the creative director of a prominent French house. I told her about a piece I was writing about synthetics for *The Times*, explaining the role that synthetics had in perfume and that most perfumes are made of synthetics today. She looked at me with honest horror. She said, *"Mais Chandler, tu casses le rêve!"* But, Chandler, you're destroying the dream! The dream being some information-free version of perfume where the stuff presumably flows purely out of a tiny magic spigot attached to a rosebush or something and is bottled by fairies with LVMH employment contracts.

I like this woman. She's serious and smart, but she shares this viewpoint with the overwhelming majority of French perfume industry people (and basically the same number of their New York counterparts), and I couldn't disagree with them more. When I repeated the comment to Frédéric Malle, he rolled his eyes and said, "They're killing themselves with this *rêve*, which in my opinion is more of a *cauchemar.*" A nightmare. For example: not only are synthetics fascinating; they're basically completely misunderstood by everyone. Including some of the pros, by the way.

Ellena and Gautier basically agreed with Malle about the vapidness of almost all perfume marketing, although cautiously. It's true that they could more easily afford to since, unlike the woman at Ducasse, they had no licensor to serve, no corporate executives to answer to, and an in-house perfumer who was part of Hermès itself. And he was a perfumer singularly capable of talking about perfume (a talent few of them have), which meant he could talk in a way that didn't freak out the consumer when he (if he) brought up the dreaded synthetics. So they were risking less. But given those caveats, they were still notably enlightened.

Gautier, for example, was truly and blessedly free of the misconception that "natural" materials equaled "good," an archaic piece of dogma ardently subscribed to by those who know nothing about perfume. A modern perfume's being made entirely of naturals is as appropriate as a skyscraper's being built of wood and thatch; modern buildings use steel, glass, and spun silicates because they are the best materials for the job. The first perfume synthetics were created in the 1800s. I once met a pompous Frenchwoman who said, "Myself, I only buy Chanel scents because they're one hundred percent natural." A stupid comment; *Chanel No. 5* is probably the best single example of the use of synthetics in the world. Its key is a molecule called an aldehyde, first synthesized in a laboratory in France in the late nineteenth century, and it's this synthetic that gives *No. 5* its volume and glittering abstract brilliance. Aldehydes are marvelous, pure powdery scents that smell metaphorically like the brilliant light from a magnesium camera flash. In fact, not only do all Chanel perfumes contain synthetic molecules; every single great scent from every perfume house is built with them.

The synthetic molecules Karanal, ethyl maltol, and a third known commercially as Calone (its chemical name is 7-methyl-2H-1,5-benzodioxepin-3(4H)-one) alone have allowed a huge amount of innovative, surprising, daring scent creations. Gautier would tell you straightforwardly that Hermès's cash cow *Eau d'Orange Verte* is 70 percent natural, and she would immediately add that, first, 70 percent is abnormally high, extraordinarily high actually; almost all perfumes on the market, including, she would stipulate, almost all Hermès perfumes, are around 20 percent natural. And second: So what. Who cares. Opposing synthetics in perfume is like opposing antibiotics in medicine. It is simply complete retrogression.

Ellena, for his part, had no more time for the antisynthetics religion than she did. But he was careful. Ellena has in fact created, for a major fashion house, a perfume that is a huge commercial success, 100 percent synthetic, and a gorgeous piece of work on every level, but he would not allow me to give its name simply because, he said with a sigh, "the public can't handle the truth." This was an argument he used often when we discussed the behind-the-scenes questions. My view was that the public wanted the truth. "In the long run," Ellena would reply, and I would agree—but his problem was getting from now to the long run. Risky.

And still he yearned for the long run, when the public *could* handle perfume as it actually is. He found it all ridiculous and attempted, whenever he was interviewed by *Le Figaro* or *L'Express*, to enlighten. Synthetics didn't merely start modern perfumery. They are modern perfumery. In 1882 Houbigant launched *Fougère Royale*, the first perfume containing a synthetic molecule—coumarin, a

deliciously sweet, chewy scent like marzipan—to become popular. From the 1500s until 1882, perfumers had just re-created
nature: "Here is a rose." From 1882, synthetics freed them to begin creating art. In 1889 Guerlain launched *Jicky* using vanillin, a
synthetic. When asked why he used synthetics, the great Aimé
Guerlain said simply, "Because they gave me an effect I couldn't
get from naturals."

To Ellena, it's a given that creating a perfume without synthetics
is like painting a picture without blues or reds. You could, but why.
Synthetics give you range, from the amazing milky lactones making
Gucci's *Rush* the ingenious piece of abstract art that it is (if Versace's
The Dreamer is the smell of silk, *Rush* is the smell of the most excellent rayon) to the gorgeous synthetic iris the perfumer Olivier Polge
created when he made *Dior Homme.* Fresh's *Mukki* and Issey Miyake's
Le Feu are both utterly, brilliantly strange because each has a drive
train that runs on hydroxy butyl thiazol, an ingenious, disorienting
synthetic that smells like milk and can give a perfume a beautiful
creamy scent. Like Toyota, Mercedes, and Ford, the Big Boys wage
fierce, constant warfare to put out the most commercially successful
new models. Françoise Donche, Givenchy's creative director, once
declaimed to me, *"Firmenich, c'est la haute couture de la molécule!"* (Firmenich is the haute couture of synthetic materials), though naturally
this is hotly contested. Sure, said a Givaudan chemist, Firmenich
had introduced Helvetolide, Romandolide, and Coranol, and they
were serious machines, but Symrise had come out with some very
strong trendsetters like Ambrocenide, Globanone, and Globalide,
IFF put Trisamber, Two-Eyed-Musk, and Florol Super on the market, and Givaudan had launched Javanol, Azurone, and Safraleine.
("At the last World Perfumery Conference," another chemist said to
me, "I think the best introductions were definitely IFF's—Florol
Super and Trisamber. Awesome.") So they're all in fierce competition over the molecular market.

The secret of Dior's *Eau Sauvage* by perfumer Edmond Roudnit-ska is Firmenich's methyl dihydrojasmonate, a molecule that smells beautifully of clean, pure light. The heart of Houbigant's 1912 cult favorite *Quelques Fleurs* is a synthetic called hydroxycitronellal. *Angel's* secret is the molecule ethyl maltol (which was isolated in 1969 and is the succulent sweet molecule you smell when you smell cotton candy). There were, of course, synthetics Ellena hated, but he hated them because they were boring, or clichéd, or inferior in perform-ance—rational reasons.

Marketers have, logically, seized on the prejudice of the "all-naturals" movement as a marketing device and promoted the idea that natural materials are always good when, of course, a low-quality natural narcissus will smell like garbage while a good syn-thetic heliotropin is an olfactory marvel, a powdery floral scent, both delicious and abstract, somehow crossed with a cloud. Then there is the universally known fact that synthetics are more likely to cause an allergic reaction. Exactly wrong. Naturals are more likely to cause allergic reactions, and for a very simple reason. To get a sandalwood smell in a perfume, Ellena could use Sandalore, a syn-thetic molecule that smells like sandalwood. It's exactly one mole-cule, $C_{14}H_{26}O$, which gave him exactly one possibility of an allergic reaction. Or he could use a natural sandalwood, which contains hundreds of molecules. Here is an extremely abbreviated GC/MS analysis of East Indian sandalwood (*Santalum album* L. from the family Santalaceae).

(*l*)-alpha-Santalol	50.00
(*Z*)-beta-Santalol	20.90
epi-beta-Santalol	4.10
(*Z*)-*trans*-alpha-Bergamotol	3.90
alpha-Santalal	2.90
cis-Lanceol	1.70

(*E*)-beta-Santalol	1.50
beta-Santalene	1.40
Spirosantalol	1.20
(*Z*)-Nuciferol	1.10
epi-beta-Santalene	0.97
alpha-Santalene	0.82
beta-Bisabolol	0.64
(*E*)-alpha-Santalol	0.56
beta-Santalal	0.56
dihydro-alpha-Santalol	0.38
ar-Curcumene	0.26
alpha-Bisabolol	0.26
beta-Curcumene	0.13
trans-alpha-Bergamotene	0.12
(*Z*)-trans-alpha-Bergamotal	0.10
complex rest	6.50
total	100

The natural creates a much greater allergy potential. If a natural has, say, one hundred molecules, the house must contend with roughly a hundred possible allergic reactions. The synthetic, just one.

Ellena would also point out that naturals pose a sourcing challenge, which creates a threat to the perfume's memory switch. For memory to trigger, a perfume has to smell the same, and everyone in the industry can remember how perfumers had been specifying natural vetiver from the Caribbean island of Montserrat when the island turned into one big volcano, blew up, and caught on fire. The entire island was evacuated, and the supply of vetiver stopped dead. Everyone had had to hurriedly switch to other sources of vetiver— Haiti, Réunion, China—(and then rush the quality controls and worry about the purity and the toxicity, and as fast as possible rebalance formulae; it had marked them all), which is why naturals also pose a much greater quality control challenge. What the all-naturals people don't realize is that synthetics are better ecologically. The sandalwood forests of India are being destroyed at a terrible rate, literally disappearing, and the price of natural sandalwood is skyrocketing (it was, not long ago, $300 a pound, then $500, then as even more Indian trees were sawed down it shot past $800 and up). Most perfumers I know refuse now to use natural sandalwood in perfumes, and their bosses support them, and it's a purely eco question. (Smugglers and the black market keep prohibited sandalwood flowing.) They don't necessarily tell Dolce & Gabbana; they just quietly put them in—lovely mixes of synthetics like Santaliff (from IFF) and single molecules like Javanol (made by Givaudan), a strong, nicely diffusive creamy/rosy sandalwood scent which saves the natural environment.

The idea that synthetics are "modern" and "American" and naturals are "classic" and "French" is completely wrong. There's no Frencher house than Guerlain, no more classic collection of perfumes, and not only were Guerlain perfumers at the forefront of the synthetics revolution with the 1889 *Jicky*, Guerlain's classic *L'Heure Bleue* (1912) derives its beauty from methyl anthranilate, its elegant *Mitsouko* (1919) uses the very elegant synthetic aldehyde C-14 (which smells deliciously of delicate, ripe peach), and the immortal *Shalimar* (1925) has quinolines. The sleek and, in my view, vastly underrated *Samsara* has the synthetic sandalwood molecule Sandalore, and just a few years ago the perfumer Maurice Roucel put a really cool molecule—*cis*-3–hexenyl salicylate (it smells, by itself, of solar-drenched suntanned air over freshly cut green grass)—into his modern classic *L'Instant de Guerlain*. Synthetics bring order, said Ellena, like a calm teacher before an unruly class, in whose presence the students relax, find a focus.

The naturals movement would like to believe synthetics are "cheap," which is the usual ideological slander. IFF, for example, has some excellent, very pricey synthetics. All scent materials have an internal and an external price; the internal is what, say, IFF will charge Vera Wang if Wang has her perfume made by IFF; the external is the price IFF will charge Givaudan for the synthetics it makes, and if Wang goes to Givaudan and the Givaudan perfumer uses an IFF molecule in Wang's juice (maybe because Wang is in love with the idea of an abstract sweet, and IFF has on the market the achingly perfect molecule for making abstract sweets), to make money Givaudan has to put its own margin on top of IFF's external price. This is why you want your research chemists working away in their labs to smelt the hottest new molecules: They're tiny cash cows, microscopic ATM machines spitting out money. For example, here are four excellent IFF synthetics:

	PRICE PER KILO	
	INTERNAL	EXTERNAL
Muscenone	€400	€900
Muscone	470	1,000
Amberiff	950	3,000
Amberketal	2,500	6–7,000

I got these prices from a former IFF evaluator who had had it with people calling his beautiful synthetics "cheap."

Some synthetics become so expensive that IFF's business managers will take them off the available palette; the price is just too high.

The final irony is that the naturals-only people don't know what the word *natural* means. There's a molecule called linalol (the trade name of the terpene alcohol $C_{10}H_{18}O$). You can chop down scarce Brazilian rosewood trees (*Aydendron rosaeodora*) from the rain forests, steam-distill the wood chippings, and then isolate linalol by fractional distillation from the essential oil. But for years we've known that you can also create this exact same molecule, much faster and less expensively, by adding acetylene to 6-methylhept-5-ene-2-one, then doing a Lindlar hydrogenation. The same molecule. Linalol is what's called a nature identical (there are thousands of them), the same molecule whether it comes from destroying the rain forests to get rosewood or from the lab. Exactly the same. Well, there's one difference between naturally derived and synthesized linalool: the synthesized molecule is pure, the natural is impure.

The "naturals-only" movement is, of course, simply a panicked reaction to a world becoming too complex and too out-of-control. But fundamentalism is always a disaster. "It seems not to occur to these people," a South American flavorist commented to me, "that a lot of naturals are actually forbidden because they are toxic. Natural coumarin from tonka beans is poisonous. Wormwood oil has high levels of thujone and is banned, as is natural nutmeg, which is loaded with myristicine. Or so many other examples. This notion that 'being natural' equals 'being healthy' . . ." He paused, looked at the ceiling. "Arsenic is natural, and it kills you."

The fear of talking about synthetics to the public is not entirely unreasonable. It is, as an absolute policy, merely irrational—a certain degree of reality injected into the conversation with the consumer would be an exceedingly healthy thing—and has been taken to an extreme that is now a religion. Combine this specific fear with the fundamental paranoia of the luxury goods industry, where nothing more than image, quixotic and perishable, accounts for the sales of one-thousand-dollar handbags, and the result is that you have this art form, perfume, an art in which polysyllabic intimidating-sounding molecules form an intricate, mercurial, perplexing clay molded by molecular sculptors called perfumers, and these artists must function inside a business of pure illusion run by executives worried about shareholders and uncomfortable with art. No wonder the perfume industry is like a sanitarium. It's the same as Hollywood.

This was why Gautier was cautious. People can understand making-of documentaries that show movie stars behind the scenes on sets working with directors. Shots of fashion designers cutting silk taffeta has an instant logic to the consumer. It's understandable.

Hydroxy butyl thiazol is not. This is why when you say "perfume formula" to marketers, they wig out. Say these words to PR people, and they act like they forgot their rabies shots. Say them to perfumers, and they are usually (but not always) slightly nervous. In fact, they are often rather wistful about their formulae, these wonderful novels they write that the marketers then assiduously hide from everyone in the world. "I wish I could talk about my formula on the record," a perfumer once said to me regretfully. She had just, finally, finished a perfume for a major American designer—we were going over it on paper—and she was extremely proud of it, but she dutifully stipulated that it was off the record. I've spent wonderful sessions with perfumers like Pierre Wargnye and Dominique Ropion and Pierre-Constantin Guéros going over formulae, the columns of molecules that represent so many hours of agonizing, careful ratios and precisely calibrated percentages. They discuss them intelligently and with passion, and it's beautiful to be walked through a molecular creation with its creator, a personal tour of the inside of the machine by the artist who built it. Say the word "formula" to a few of the more paranoid ones, and they react like chickens on speed. They believe the formulae need to be kept absolutely secret.

But with the fundamental irony that the fervor with which the houses keep their formulae secret is actually inversely proportionate to the importance of doing so, which is either zero or near zero. It's tough to say exactly. The unspoken fact is that all formulae are public now, at least to everyone who owns the right $50,000 worth of molecular analytical machinery (which is all the Big Boys and any number of independent laboratories), but the industry is terrified to admit it for a rather simple, if not particularly intelligent, reason.

For historical reasons having to do with how medieval perfumers made their money, Dior does not own the formula to a single Dior perfume. Lancôme has never seen the formula for *Magie*, Ralph Lauren and Estée Lauder could not tell the formula of *Romance* from that of *Pleasures* because they've never seen them. *J'adore* was created by the perfumer Calice Becker at Quest, and Quest, not Dior, owns the formula for *J'adore*. When Givaudan bought Quest, Givaudan—not Dior—then owned the formula. Symrise owns the blueprint to Givenchy's *Amarige*, IFF owns the plans to Donna Karan's *Cashmere Mist*, and Firmenich owns the wiring blueprint to Armani's *Acqua di Giò Pour Homme*. Giorgio Armani has never seen his own formula.

When Ellena made *Un Jardin en Méditerranée* for Hermès, he was an employee of Symrise, so Symrise owns the formula to that Hermès perfume. In fact neither Véronique Gautier nor any president of Parfums Hermès before her (nor, incidentally, Jean-Louis Dumas) had ever seen an Hermès formula, not for *Calèche* (which was owned by Mane, which held it close in its paranoid grip), not for *Equipage*, created and guarded by the perfumers Guy Robert and Jean-Louis Sieuzac.

Ellena's *Nil* was thus a revolution for the house: It would be the first Hermès formula that Hermès would own.

The convention surrounding formulae is due simply to the fact that medieval perfumers guaranteed the faithfulness of their clients by keeping their formulae secret. Only your perfumer could recreate your perfume. So Givaudan and Firmenich and IFF, the descendants of these medieval businessmen, do the same.

The problem was that in 1952 A. T. James and J. P. Martin developed a machine called the gas chromatograph. This was an ingenious device that separates individual molecules on an invisible conveyor belt of gas and IDs each of them as they float by. It tells you what molecules are in the material before you, unlocks the juice, and hands you a printout like a grocery store cash register spitting out a paper listing of the items sitting inside your plastic bags at the checkout. (A little more complicated, but still.) And from that instant, secrecy began a slide to uselessness. The gas chromatograph and its brother, the mass spectrometer, with which it is used in tandem, do molecular analysis and, with a really good molecule jockey reading the results, and with enough time to interpret what the machines spit out, give you basically 100 percent of any formula. Everyone in the industry knows that the first twenty bottles of IFF's perfumes are bought by IFF's competitors, who take them back to labs where the techs "shoot" the juice (that's the verb they use; "We just shot the new Gucci masculine, totally blew me away how much Isobutavan they put in there!"), the machines crunch it, the techs eyeball the molecules, fill in the gaps with a little wizardry, do a polish, and send the formula directly to the hard drives of the executives. (Perfumer François Demachy, by contrast, told me that in his extremely conservative view, you could only piece together about 90 percent of a formula with the machines. Perhaps, but this is contradicted by every perfume lab technician I've ever talked to.) The IFF execs read the Givaudan formulae, the Givaudan execs read the Firmenich formulae, and everyone gets together at industry functions for cocktails and cheerfully doesn't talk about it.

It's not hard. The truth has never been a particularly significant ingredient in the industry in so many ways. I got bumped up to a rare business class seat on an American flight from Paris to New York and met in the seat next to me one of the industry's legendary figures. She spent the flight telling me stories. One was about a meeting she had attended in the vast Manhattan conference room of one of the biggest French luxury brands. A French perfumer was presenting the brand's executives their new scent, which she'd made for them. They were launching it next month, millions of dollars, media, a major deal. With great flair they brought in the *touches* sprayed with the scent, the suits sent the thing around the room on its little paper carriers, and everyone smelled deeply, pronouncing it great, *genius*, exclaiming about "the notes" of this and that. This went on. After a while one young guy screwed up his courage and said, "Uh, I just don't smell anything." Everyone froze. Dead silence. A few people cleared their throats, then started backtracking. The perfumer looked rapidly from one to the other, grabbed a scent strip, smelled it. Her face went opaque and she rushed off-stage and grabbed the bottle. Her assistant had screwed up; the bottle she'd sprayed from was a *factice*, one of the mock-ups they make for photography. It had been filled with colored tap water.

IFF is still pretending that if it keeps the formula secret, Estée Lauder can't take *Beautiful* away and have Givaudan manufacture it for less. (This would quite simply involve a Givaudan perfumer quietly taking out, say, an expensive natural Sicilian bergamot and substituting in some cheaper Brazilian stuff, then rebalancing the formula, which now costs two euros less per kilo, instantly kicking up Lauder's profit margin. Lauder would hope no one would notice, and in many cases—particularly if the perfumer has talent—they wouldn't. Note that I'm using IFF, Lauder, and Givaudan here purely as theoretic proxies.) What this means is that these are twenty-first-century companies operating under medieval business practices, the emperor has no clothes, everyone knows it, and no one will admit it.

Ellena was very much of two minds about this. He was punctilious about the rigorous secrecy of his formulae, a firm believer in and emotional adherent to this tenet of the perfume industry creed. This despite the fact that if you asked him, most of the time he would also tell you quite directly that he recognized the secrecy to be bluff. If you asked him at still other times, you could get a different answer. It depended on how he chose to see it at that moment. It was a small cognitive dissonance, but he argued some of it away and some of it he just shrugged his shoulders and accepted.

He did have one intellectual counter: His argument was that you *could* keep formulae secret because the gas chromatograph could be fought and, he would tell you, he had figured out how to fight it. He argued—others made this claim as well—that he and a few of the truly talented perfumers had become adept cryptographers, loading red herrings into their formulae, secret codes to screw up the analytical machines.

Essentially the secret codes were naturals, and the naturals were used as codes because they were so complicated. Put a synthetic in your formula, and you'd have added exactly one molecule. Shoot the juice, and on the readout the lab tech would see, for example:

dihydrojasmonate. (Very simple.)

But naturals are phenomenally complex. Rose absolutes can contain a thousand molecules. Put in a bit of rose, and on the lab tech's readout he'll see:

neryl acetate, α-terpineol, n-heptadecane, geranyl acetate, citronellol, nerol, n-octadecane, geraniol, benzyl alcohol, nonadecane$+1$-nonadecene, phenylethyl alcohol, methyl eugenol, eugenol, farnesol ... (And this is just the tiniest proportion of the immense list.)

Add naturals, and you multiplied the complexity logarithmically: The molecules would go on for pages. How did Ellena use this simple fact to encode his formulae, his trick to confuse the machines? Simple: Slip subperceptible doses of naturals into the perfumes. He could add a tiny amount of ylang-ylang essence from the Comoros, and he didn't change the smell of the perfume at all—for the consumer, it might as well not be there—but the mass spec picked it up, and the machine saw hundreds upon hundreds of molecules flooding through, all of which it faithfully reported. Do that a few times, a tiny bit of jasmine, a tiny bit of violet leaf, molecules piled on top of molecules spilling onto molecules, and the mess would be spectacular. It was the same principle of packing into a single missile ten fake nuclear warheads along with one real one: Midflight the missile spat the warheads into the air, and the radar registering the rapidly incoming blips had no idea which of the ten was actually armed.

But the counter to his counter, which a technical perfumer gave me when I asked him about coding, was: "That's fine. He can encode his formulae by adding red herrings, but the problem is, first—and Jean-Claude may or may not acknowledge this, but that's his choice—that the machines are getting smarter and smarter. Second, if you put in a complex rose at a subolfactory dose only to confuse the machine, by definition the amounts are tiny. You can follow those minuscule percentages. It slows me down a little, that's all. Trace amounts of things that I can pretty much tell they're only in there as decoys?" He shrugged. "It's fine in theory, but is Jean-Claude such a wizard that he can figure out how to permanently throw both me and the mass spec off the trail? Perhaps. He's smart. But I doubt it. People like to say, 'Oh, I can fool the machines,' but I haven't seen it much. IFF can code a Dior fragrance with trace naturals, but Dior can hire me, and I can eyeball the readouts, figure out which part is coded junk, adjust for the background noise, and simply wipe it out."

If it is essentially impossible to keep them secret anymore, there is a second, complementary reason for the fear of revealing formulae. It's another bizarre legal anachronism of the business: No one has contracts. One of the Big Boys' marketing guys once said to me flatly, "We own one of Givenchy's juices; we've made it for several years. But because our operational practices are so anachronistic we actually have no legal contract with Givenchy—or, more precisely, with LVMH, which owns Givenchy—that protects our exclusive legal right to manufacture this formula. Instead we just keep it secret. Except that we and they and everyone know that everyone has shot our juice and that everyone else has the formula for this perfume, or at least that they've got 99 percent of it.

"If we really wanted to protect ourselves, we'd have a contractual relationship saying only we can legally produce this formula, and Givenchy can only buy it through us, but no one on our side has the balls to ask Givenchy and Lauder and L'Oréal for that. We just pretend everything actually makes sense and hope they will too." He turned his hands palms up and shook them, once. "We got nothing. Fucked-up business."

The problem is actually worse than he described it. Even if L'Oréal gave Firmenich (say) such a contract for an Armani scent, were L'Oréal ever to decide to have the juice made cheaper by Takasago, it could have a Takasago perfumer just change .1 percent of the formula, and were Firmenich to sue—which it never in a million years would—L'Oréal would simply say to the judge, "Hey, the formulae are different." The change would be done in such a way that the consumer couldn't *smell* the difference, the judge would back L'Oréal, and L'Oréal would sell the cheaper Takasago-made formula and make more money. As one of them said to me, "Moral of the story: Intellectual ownership doesn't protect squat." This is one reason they walk on eggshells.

They shoot the new ones pretty quickly these days, more or less the moment they hit the market. One day I was in a perfumer's office discussing the use of a particular synthetic I like. The funny thing about it is that this molecule manages to have a different effect in different perfumes, and the perfumer and I were looking at five formulae that contained it, comparing percentages in each. He mentioned that another perfume, just launched, also contained a pretty healthy dose of the stuff, "but they haven't shot it yet," he said. He frowned. "Wait." Hit a few keys, peered at the screen. "Yeah," he said, "it just launched on Friday; they'll probably shoot it Wednesday; I'll let you know."

The perfumers each choose how they will behave with me. I was looking at formulae with one, and she took a piece of paper and carefully covered her screen to hide the internal price column. But then I spent an afternoon with another perfumer at the same company, and he just shoved the screen in my direction and we started working. He said, "Obviously you're not going to screw me or my company." And obviously I'm not. I'm neither stupid nor malicious. There is no point to that.

Because of all of this, when I started this beat for *The Times* and began to report on formulae and the synthetics in them, though I didn't realize it immediately, the industry viewed me as a threat. One incident illustrated it for me. I was establishing with a Big Boy what we were calling "guidelines," i.e., good playground behavior, what would be on and off the record, etc. They definitely wanted me to write about their perfumes, and their work (after all, their job is creating the formulae and making the scents, and they're justifiably proud of them; the brands just think up cool names, stick on pretty labels, and market the stuff). And yet at the same time the whole thing made them profoundly uneasy because if Armani freaked because I wrote that the latest perfume they'd made for Armani had this or that molecule in it, and if one of their employees had told me about the molecule, Armani could punish them by not sending them its next brief proposal for the next Armani perfume. So I was busy reassuring them and we were going back and forth discussing fact-checking practices and defining journalistic parameters, and their corporate immunological system was busy trying to figure out how to react to me: Was I self or nonself?

They were sending e-mails about this to five different people in various departments. The communications person in New York was negotiating with me, and suddenly I got an e-mail that came from one of the execs in Paris, addressed to her with everyone else cc'ed: "Good job, Carla, keep up, Chandler is difficult to control . . . let's all be very careful with him."

I've inadvertently sent e-mails to the wrong person as well. I said so in my reply, and I also said that I was glad to have gotten his e-mail because now we could speak frankly. (A few weeks earlier, I'd gotten a late e-mail from an industry chemist. "Sorry for the delay here," he'd written, "I'm trying to figure out how dangerous you are.") Why, I asked, did they need to be so careful? His response came two days later. "Dear Chandler, Nothing personal, be assured, just the usual issue of respecting our customers' communication. We have had issues in the past, believe me, with customers reacting negatively at what we have been saying to journalists . . . I remember a lot of 'dramas' . . . Therefore we avoid to speak about our customers' products, by experience."

I had to learn that were I to write a piece with a Big Boy's participation about, say, a Ralph Lauren perfume, and if the slightest detail appeared that the Ralph Lauren PR department hadn't itself vetted, such as, say, that amyl allyl glycolate played a significant role in the new Lauren launch, then Ralph Lauren would go ballistic. This despite the fact that the innovative use of this molecule is (oh yeah) both true and what is interesting and substantive about the perfume. But the perfume industry shuns substance like the plague, and they are saddled with a lie of their own creation that undergirds perfume's most fundamental marketing platform. The Ralph Lauren PR people want you to write that Ralph "created" his perfume. You, however, know the truth, that perfumers created it, that Lauren's creatives managed the process, and he himself smelled a few iterations, maybe, in some meeting at the very end of the process and said "Yeah fine," and that was it. So—forget even the synthetics and formulae for a moment—just the fact that I wrote about *the perfumers* freaked them out. It freaked out the perfumers too, a little, which seemed crazy: The perfumers were the story, as far as I was concerned.

I first met one of SJP's perfumers, Laurent Le Guernec, at a large business dinner a few nights after my piece on Parker came out in *The Times*. Some of his friends had told him, "Hey, Laurent, you're in *The New York Times!*" He hadn't believed it (*"Allez!* My name?"), he'd gone and looked and been amazed to see it, and he came up and shook my hand and said he appreciated it. I was amazed he was amazed. It made me angry, not at Le Guernec but at the industry. Here's this guy, he's the *creator*. And they keep him in a cage, in the back, in the dark.

In the "dream" version of perfume, marketers tell the public that perfumes have "notes of caramel and blueberry," which simply means, since there are no natural caramel or blueberry perfumery raw materials (it's neither technically possible nor financially viable to distill them), that the perfumers have just created these scents (perfumers call them accords, not notes, which is a term for public consumption). You can create the scent of caramel with 3-hydroxy-4,5-dimethyl-2($5H$)-furanone. If you take that molecule and add a small amount of ethyl butyrate, ethyl valerate, and phenethyl acetate, you get a nice fresh garden berry that would work great in an Escada launch. God forbid the public knew it.

Explaining a jet engine or the wing of a 787 doesn't destroy the awesome beauty of flight. It doesn't break the dream. It does the opposite. The more you understand of science, the more you marvel at the magic of reality, and creating the dream is not the same as perpetuating ignorance. It is the opposite: taking people inside, letting them see behind the scenes, showing them how it all works. To the degree to which its public discourse aligns with the truth about the construction of its perfumes, Estée Lauder is always on surer, safer, more solid ground. This is, pretty much, the fundamental political observation of the twentieth century; it is one of the more obvious economic lessons drawn from ideological, antimarket socialist economies where both economic forces and the public relations surrounding them were divorced from the reality of consumer instincts. Lauder's old public relations policy, in which the perfumer was never to be mentioned and Mrs. Lauder was presented in some vague, inchoate way as sitting in her kitchen pouring raspberry ketone into dihydrojasmonate, is from a different era. The paradigm is antiquated.

I would suggest that it is also commercially ineffective. In fact, probably counterproductive. Perfume sales have been flat for years; the perfume industry has ferociously resisted allowing consumers behind the curtain. I have seen their formulae. Lauder formulae, Hilfiger formulae. The point is not formulae. The point is that decades ago every other industry started taking their consumers behind the curtain, and we're fascinated by everything from how Warner Brothers makes superheroes fly on screens to how Toyota makes its cars, Frank Gehry his buildings, presidential election machines their candidates. The perfume industry alone remains walled off like a gulag.

Millions are fascinated by the process by which designers like Todd Oldham cut, sew, design, and agonize their fall collections into existence, but the great creative minds at Yves Saint Laurent and Jean Paul Gaultier and Dior, with the collective brilliance of a single mollusk at low tide, have intuited that with perfume—No. Here is an industry suffocating itself on the most immense pile of public relations shit human civilization has ever produced, a literal mountain of verbiage about "the noble materials, symbol of eternal feminine beauty, addictive notes of Cocoa Puffs, she can't wait to taste him like a Hershey's kiss, Cleopatra wore this, it has notes of distilled wild all-natural Martian fungus harvested by French virgins on the third moon of Pluto." The lies pile up on other lies, they generate a poisoned river of vapid crap the marketers try to pass off as "information," and the brands have no clue that their public relations approach is about fifty years out of date. Reading anything they put out on their perfumes is like reading a combination of Kafka, only less creative, and *Pravda* circa 1985. Zero interest. There is almost no recognition that the enforced lack of knowledge—this gaping void of nothingness about what their products actually are, who makes them, and what's in the things— is creating boredom and disinterest. The perfume industry is choking itself to death on its vacuum.

Look at the sales figures.

The taboos are the problem. One of these is against publishing perfume formulae. So here is a perfume formula.

My friends send me formulae; I keep them carefully in a subfolder on my hard drive. I have the formulae for Dolce & Gabbana *Light Blue*, Hermès's *Ambre Narguilé*, Ralph Lauren's *Romance*, Estée Lauder's *Beyond Paradise*, and so on. I have the formula for Cartier's *Déclaration* because an industry friend heard I was writing about

Ellena—it's his juice—and sent it to me *"afin que tu puisses connaître sa patte,"* so you can get to know his style. Some I don't have anymore because they were sent to me on the condition that I look at them for my personal interest and delete them. One of these was for *Un Jardin sur le Nil.* A friend sent it to me just after the launch but asked me not to keep it, so I read through it—a few surprises, but most of it I knew from what Ellena himself had told me—and deleted it.

So this formula. It is for a hugely successful commercial luxury perfume. It's a feminine, it's from a house named for an iconic French designer, and it is in my view one of the most beautiful perfumes ever made. I got it from a friend who shot the perfume in a lab, so it is not the original, but it is probably 98 percent+ of the original on a molecule-to-molecule comparison, and I'd bet 100 percent of the original in terms of smell perception. Technically it is expertly done, with excellent diffusion and excellent persistence on skin, seamlessly built. As for the aesthetics, it is a wonderfully ripe, fruity, full scent with great volume, smooth and strong as a swimmer's naked back and as self-assured and direct as the gaze of an African woman.

The thirty-three materials—more or less the average number—are listed alphabetically, and volume is indicated in parts per formula totaling 100.

allyl amyl glycolate 10%	0.40
Ambrettolide	1.10
bergamot Italie	3.00
cardamome Guatemala	0.30
carvone Laevo 10%	0.10
cassis base 345B	9.10
cis 3 hexenol 10%	1.30
cis—3—hexenyl acetate 10%	0.40
damascone beta 10%	0.50
dimethyl benzyl carbinyl butyrate	0.30

dipropylene glycol	18.5
ethyl aceto acetate	0.90
ethylene brassylate	30.4
fleur d'oranger F175SAB	0.20
geraniol	0.50
geranium bourbon	0.40
geranyl acetone 10%	0.40
grapefruit	0.50
Hedione high cis	8.90
helional	0.90
β-ionone	1.00
lemon Italian	0.90
linalol	1.30
linalyl acetate	1.20
magnolan	3.30
methyl pamplemousse	0.50
mousse synthétique 10%	0.20
nerolidol	5.60
phenoxy ethyl iso butyrate	0.30
phenyl ethyl alcool	2.00
reseda body 10%	0.10
Sandiff	5.30
tuberose base	0.20
	100.00

I N PARIS, HÉLÈNE Dubrule, as head of marketing for Hermès's perfumes—both those extant and those Ellena would be creating—was responsible most centrally for how the house's perfumes would be sold.

One of the first meetings I sat in on at Hermès was, a bit quixotically, the most sensitive meeting possible. They'd only just started to get used to me. Dubrule ran the meeting. Its point was the elucidation of the broadest strategy for Parfums Hermès.

Hermès itself is oddly fragile. It is far more bothered by the difficult commercial environment in which perfume is sold than, for example, Ralph Lauren (whose perennial commercialness is robust) or Missoni (which can cruise cushioned by a certain Italian loyalism). Hermès is French, and thus it is brittle, difficult to change without breaking, a quality that counts among the greatest French weaknesses. This was what Dubrule was facing. That said, perfume poses a problem for any serious luxury house. Edith Touati, Hermès's director of international marketing from 1984 to 1990, had faced the weird perfume problem, which boiled down to a question of brand control. During her tenure, she had opened boutiques all over the world, and, she said, the only product that made her uncomfortable was perfume, for one simple reason. The house's silk ties, its burnished shoes, its shirts and saddles, its dinnerware and watches, all its products were sold in Hermès boutiques. All except the perfumes.

"It's difficult to talk about Hermès fragrances," Touati said in a practical, straightforward way. "Their products are sold only in their stores while their perfumes are sold in perfumeries: Saks, Sephora, Barneys, *other people's* places, and then"—she grimaced—"the discount stores with fluorescent lighting that get hold of some bottles illegally and stack them on metal shelves next to aftershaves. *C'est un produit Hermès qui est sorti de l'univers Hermès.* Perfume is an Hermès product taken out of the Hermès universe. Which doesn't always put it in a favorable position." There are fourteen Hermès stores in the United States, thirty in France, and thirty-six in Japan, and each is an immaculate temple. But Hermès perfumes are sold in thousands of crappy little stores presided over by fat men with mustaches.

Touati recognized that all the houses could say the same thing—"Imagine Comme des Garçons in some *pharmacie* (drugstore)?" she said—but argued that Hermès has a stronger identity than almost any other house. "It's a very coherent, elitist brand, and *that* is precisely the problem: Perfume is no longer an elitist product, and the distribution system is such that it's sold in other people's environments, which Hermès does not and cannot control. Perfume is thus *quotidien,* everyday."

She paused, noted pointedly: "Everyone, every year, asks the same question: 'Why is there no Vuitton perfume? Why doesn't Vuitton jump in?' The answer is obvious. It's because there would then be Vuitton products in some store that's going to treat them like any other product, store them on some *shelf* somewhere, and put them on sale at 50 percent off, and they'll have no control over their own products. And this is a brand that never discounts any product, a brand as obsessed with control as one can be. And you really think Vuitton should do this?" She paused again, then said, "*Non, je ne pense pas.*" No, I don't think so. "Clearly they don't either."

Dubrule was as conscious as anyone of what had made the List—as it is known in the industry—and what had not. The perfume industry is just as obssessed with the best-seller lineup as Hollywood is with the international box-office scores, and the executives and perfumers pore over the lists like shamans over runes, reading the signs, trying to divine trends. Dubrule knew the previous year's lists—2003, the masculines and feminines (in an indication of how outmoded the assessments are, there is no list for the most interesting category of perfume, mixed scents)—but what was one supposed to glean from them?

The bestselling feminines in the United States in 2003, when the U.S. fragrance industry was worth around $6 billion, were the following:

1. *Beautiful*	Estée Lauder
2. *Happy*	Clinique, a brand created and owned by Estée Lauder
3. *Pleasures*	Estée Lauder
4. *Romance*	Ralph Lauren, licensed to L'Oréal; L'Oréal paid Lauren a licensing fee
5. *Cashmere Mist*	Donna Karan, whose license was owned by Estée Lauder
6. *Chanel No. 5*	Chanel, which was independently owned by Alain Wertheimer
7. *Trésor*	Lancôme, a subsidiary of L'Oréal

8. *Beyond Paradise*	Estée Lauder
9. *Glow*	Jennifer Lopez, whose license was owned by Coty
10. *Chance*	Chanel
11. *Miracle*	Lancôme (L'Oréal)
12. *Happy Heart*	Clinique (Estée Lauder)
13. *Angel*	Thierry Mugler, whose license was owned by Clarins
14. *Ralph*	Ralph Lauren (L'Oréal)
15. *Vera Wang*	Wang's license was owned by Unilever
16. *Aromatics Elixir*	Clinique (Estée Lauder)
17. *Coco Mademoiselle*	Chanel
18. *White Linen*	Estée Lauder
19. *Eternity*	Calvin Klein, whose license was owned by Unilever
20. *J'adore*	Dior, which was owned by LVMH

In other words, of the top twenty, Estée Lauder owned eight, L'Oréal owned four, and Chanel, three. Unilever owned two. Coty owned one, and LVMH owned one. This lineup had changed little since the previous year, 2002, when the top four had been *Beautiful, Happy, Pleasures, Romance,* all identical to the top four in 2003 and in the identical order. In fifth place was *Chanel No. 5. Cashmere Mist* had been seventh.

The List is made up of sales data from various sources, the sources change and are never complete, and the full picture is never entirely clear. This means that neither the Big Boys, who make the perfumes, nor the brands, such as Hermès, ever get total information. Still, the List is, generally, an accurate picture of the field. NPD is the company that collects U.S. industry sales data, organizes and analyzes it, and then sells it back to the industry. (Every player buys NPD's products, and the bigger they are, the more data they buy.) As soon as the bar code on a box of perfume is swiped, it is automatically added to the NPD stats, which at the time covered around 85 percent of stores, though for complicated political reasons not Saks 5th Avenue, which meant that when Narciso Rodriguez's eponymous scent became a big hit at Saks it didn't appear on the List because it was a Saks exclusive. Until 2002, NPD didn't cover Sephora, which made the List look very different after 2002.

NPD recorded that the best Hermès performer in 2003 in (and only in) U.S. department stores (NPD doesn't collect data from, among others, Hermès boutiques) was *24 Faubourg* at $613,945 for 6,513 units (bottles of perfume). As such, *24 Faubourg* came in at number 217 on the List, nothing to write home about, to put it mildly. This was the year before Hermès took on Ellena.

In 2002, *24 Faubourg* had been $451,027; in 2000, virtually nonexistent commercially—at number 301 on the List—at $7,430. Hermès was playing around the number 300 mark: In 2003, *Calèche* was number 293 ($164,758), *Hiris* was number 330, and *Rouge Hermès* was number 316. The Hugo Boss marketers wouldn't have gotten out of bed for numbers like these, but on the other hand the Hugo Boss perfume creatives produce nerve gas. Hermès was doing slightly better with the fragrances it marketed to men: *Bel Ami* came in at number 164 with $105,318, *Equipage* was number 179 ($42,985), and *Rocabar* was number 177 ($55,306). (Notice that a men's number 177 can be $55K, where a women's number 293 comes in at $164K.)

Chanel No. 5 sales in 2003 were reputed to be around €180 million, "and," said an executive at Chanel's most direct rival, "this perfume is *eighty years old!* It's unbelievable! It's not a fragrance; it's a goddamn cultural monument, like Coke. It's the reference."

Hermès's numbers were nowhere near this. Dubrule discussed them. Everyone else discussed them too. I was having lunch in a restaurant on the place de Barcelone with a woman in a smart Chanel suit and red lacquer nails who had Dior's and

Kenzo's and Issey Miyake's numbers in her head, and when she could not recall those from Hermès, she put down her salad fork, picked up her cell phone, and *"Didier,"* she said breezily, *"tu peux me chercher le chiffre d'affaires des parfums Hermès?"* (Could he grab her Hermès's grosses for its perfumes?) She ate a radish for a moment, gazing at the sky with the phone between an ear and a shoulder. Then he was back. *"Oui,"* she said, then *"Cinquante-cinq millions d'euros dans la parfumerie,"* she said, and then, *"tandis que les gros revenues sont d'un milliard trois cents."* Hermès made €55 million annually in perfume on a €1.3 billion gross. *"D'acc, merci."* She hung up, shrugging. *"Ça participe à l'image."* The perfumes add to the image.

One thing about the List is that it's quite different depending on the country. In 2003 *Chanel No. 5* was number one in Great Britain, number three in Spain, number two in Italy and Germany, and number one in France. (It was also number one on the European list, due mostly to the huge weight—the sheer number of bottles rung up—that the French consumer gives the French list. In France, 170,000 bottles of perfume are sold every day.) Contrary to the U.S. lineup, the list of the 2003 bestselling feminines in France looked like this.

1. *Chanel No. 5*	Chanel
2. *Angel*	Thierry Mugler (Clarins)
3. *J'adore*	Dior (LVMH)
4. *Flower by Kenzo*	Kenzo, which was a brand owned by LVMH
5. *Shalimar*	Guerlain, which LVMH had bought in 1994
6. *Chance*	Chanel
7. *Coco Mademoiselle*	Chanel
8. *Allure*	Chanel
9. *Opium*	Yves Saint Laurent, which was

	owned by Gucci, which was owned by Pinault-Printemps-Redoute
10. *Jean Paul Gaultier*	The Gaultier license was owned by a French company called BPI, which was actually Japanese, a carefully disguised subsidiary of Shiseido
11. *Paris*	Yves Saint Laurent (PPR)
12. *L'Air du Temps*	Nina Ricci, whose perfume license was owned by the Spanish licensee Puig, which created its perfume.
13. *Trésor*	Lancôme (L'Oréal)
14. *Eau de Rochas*	Wella of Germany owned the Rochas license
15. *Coco*	Chanel
16. *L'Instant de Guerlain*	Guerlain (LVMH)
17. *Lolita Lempicka*	Pacific Europe, a Korean licensee
18. *Aromatics Elixir*	Clinique (Estée Lauder)
19. *Ô de Lancôme*	owned by L'Oréal
20. *Dior Addict*	Dior, owned by LVMH

Of the top twenty, Chanel owned five, and so did LVMH. PPR and L'Oréal each owned two. Estée Lauder owned one, at the bottom, and its number one bestseller in the United States, *Beautiful,* didn't even appear on the French list, nor its number three, *Pleasures,* nor its number eight, *Beyond Paradise.* Neither *Vera Wang* nor *Glow* made the list in France; nor did either Ralph Lauren's bestseller or Donna Karan's *Cashmere Mist* or the Calvin Klein. Even one of the French perfumes that made the U.S. list, *Miracle,* did not place in France. All in all, of the top twenty, twelve scents on the American list didn't appear on the French list, all of them American; thirteen of the French best sellers didn't appear on the U.S. list, and of those, twelve were French.

The number one scent in Germany was *Jil Sander Sun,* by a

German designer, which appeared on none of the other lists. In Spain *Anaïs Anaïs, Noa,* and *Emporio Armani* made it; *Noa* showed up in no other countries. Jennifer Lopez's *Glow* made the list in Germany (at number twenty) but not in Italy or Spain. Lauder, which completely dominated the American list, managed to get two in Spain—*Pleasures* and *Happy*—and four in Great Britain but none in Italy or Germany.

As for the masculines, the 2003 U.S. list looked like this.

1. *Acqua di Giò Homme*	L'Oréal owned Armani's license and made his perfumes
2. *Polo Blue*	L'Oréal owned Ralph Lauren's license and made his perfumes
3. *Eternity for Men*	Unilever owned Calvin Klein's license and made his perfumes
4. *Obsession for Men*	ditto
5. *Armani Mania*	Armani (L'Oréal)
6. *Polo*	Ralph Lauren (L'Oréal)
7. *Curve for Men*	Liz Claiborne, which was a licensee and created its own perfumes
8. *Romance for Men*	L'Oréal for Ralph Lauren
9. *Chrome*	Clarins owned Azzaro's perfume license
10. *Tommy*	Hilfiger's license was owned by Estée Lauder
11. *Polo Sport*	Ralph Lauren (L'Oréal)
12. *Kenneth Cole NY Black*	Coty owned the Kenneth Cole license
13. *Cool Water*	Davidoff, which was a brand owned by Coty
14. *Drakkar Noir*	Coty owned the Guy Laroche brand

15. *Le Male*	Jean Paul Gaultier's license was owned by BPI, which was owned by Shiseido
16. *Pleasures for Men*	Estée Lauder
17. *Intuition for Men*	Estée Lauder
18. *L'Eau d'Issey Homme*	Issey Miyake, owned by BPI, owned by Shiseido
19. *Aramis*	A brand owned by Estée Lauder
20. *Lacoste Pour Homme*	Procter & Gamble

L'Oréal six, Lauder four, BPI two. LVMH, zero. It was the second year in a row at #1 for *Acqua di Giò Homme,* a juice by perfumer Francis Kurkdjian that for L'Oréal, which had the Armani license, was essentially a license to print money. Other than American men having purchased enough insecticide to get *Hugo* from Boss, *Aramis,* and *Chrome* on the list and the appearance of Kenneth Cole (weed killer), the previous year revealed nothing except (1) there were no Hermès perfumes and (2) American men had, overall, bad taste. Interestingly, French men had even worse taste. The 2003 French masculines list looked like this:

1. *Le Male*	Jean Paul Gaultier, BPI, Shiseido
2. *Eau Sauvage*	Dior (LVMH)
3. *Azzaro Pour Homme*	Azzaro was owned by Clarins
4. *Allure Homme*	Chanel
5. *Fahrenheit*	Dior (LVMH)
6. *Chrome*	Azzaro (Clarins)
7. *Boss Hugo Boss*	Hugo Boss, which was owned by Procter & Gamble
8. *Habit Rouge*	Guerlain (LVMH)
9. *Hugo*	P&G, creating perfumes for Hugo Boss
10. *A*Men*	Clarins for Mugler

11. *1881*	Nino Cerruti's license was owned by Unilever
12. *Acqua di Giò Homme*	L'Oréal for Armani
13. *Biotherm Man*	Biotherm was owned by L'Oréal
14. *XS*	Paco Rabanne's license was owned by the Spanish Puig
15. *Egoïste*	Chanel
16. *Kenzo Pour Homme*	Kenzo was owned by LVMH
17. *Boss In Motion*	P&G for Hugo Boss
18. *Kouros*	Yves Saint Laurent, whose license was owned by YSL Beauté
19. *L'Eau d'Issey Homme*	Miyake (BPI) (Shiseido)
20. *Higher*	Dior (LVMH)

Lauder zero. Procter & Gamble on the other hand had sold enough disinfectant to Frenchmen to give it three placings.

The feminine lists—particularly the French lists—represented a certain happy consensus on some quite decent perfumes: *Jean Paul Gaultier* and *Paris* are hands down gorgeous, it's always nice to see *L'Air du Temps,* and Jennifer Lopez's *Glow* is, in fact, a very well designed commercial perfume. *Beyond Paradise* is an admirably daring conceptual device that succeeds. *J'adore* shows what modern luxury smells like; *Flower by Kenzo* is a delightful contemporary floral. *Coco's* coming in at fifteen showed that Frenchwomen could still wear a heavyweight French classic and *L'Instant de Guerlain's* at sixteen that they could also understand a light luminous French modern. There were also a few slips. It is time to recognize that *Trésor* smells like celery, and not in a good way, and *Eternity's* time has probably come and gone; *Chance* is, by Chanel's standards, merely nice and not particularly interesting, innovative, or deep. That said, it was encouraging to see that perfumes with the exquisite youthful sophistication of *Coco Mademoiselle,* the daring of *Angel,* and the full-throated design of *White Linen* and *Shalimar* could still top the charts and pull in the benjamins.

The List means financial reward, but it does not necessarily mean profit. Take Calice Becker's *J'adore*, the blockbuster she created in 1999 for Dior. In just a few years, *J'adore* had topped a breathtaking €130 million, and apparently it has earned a quarter of a billion dollars—this is a single perfume—for Dior and is still hauling in grosses by the bucket. (It's made money as well for Givaudan, the Big Boy that is Becker's company and that produces *J'adore*'s juice for Dior: mixes the raw materials in the formula, manufactures the solution in alcohol at X percent, chills, filters, colors, and ships this juice directly to the filler, who bottles it. I was told that Givaudan has made on *J'adore* perhaps €15 million, which if it were true would give you an idea of the difference in markup from producer to brand and from brand to consumer, although I wouldn't bet anything on this number.) What is certain, by contrast, is that the vast, infamous markups on perfume are exaggerated, not in specific but in ultimate terms. Specifically the margins, measured strictly on the cost of the juice in that bottle, are large. Scent makers like Givaudan generally put a margin of 3 or 4 on their formulae; that is, if they have a $10 RMC (raw material cost) per kilo for some given stuff, they'll charge Lauder $40 per kilo for it, though I've heard of margins from 2 to 6, depending on the material, the scent maker, the brand, et cetera. (Incidentally, "multiple" and not "margin" is, to be technical, the correct word to use, "although," as one industry guy responded when I asked him about it, "no one says 'multiple.' It's like you have to misuse the language to show you're an insider." He headed right back to the money question, forcefully: "Listen, when you consider our massive R&D costs, 'K? Our net profits are not all that glamorous.")

Obviously when the brand (Dior, say) retails the raw materials to you, you're paying the RMC plus Givaudan's margin plus Dior's margin plus Bloomingdale's margin, but again—not all that glamorous: Dior's overhead costs have become so great, and marketing, advertising, shipping, commissions, splits, kickbacks of various dismal kinds so significant, that in the end profits in the perfume industry come back toward what insurance companies tend to make. This is actually quite decent (healthy insurance companies clean up), but still, it is a tough, risky, brutal, expensive business, one proceeds into it well advised, and one survives in it through constant struggle. The executives at Dior and Givenchy and Armani have wonderful dinners at Per Se and Georges, but they also have sleepless nights.

When it works, it works like nothing else. Estée Lauder introduced *Youth-Dew* (made by IFF), and the perfume became the company's engine. Business went from, roughly, $300 a week in some stores to $5,000, unheard of in 1953. By the time Leonard Lauder joined his mother's company in 1958, total sales were about $800,000 a year, much of that from *Youth-Dew.*

Also, frankly, smell is so powerful that given half a chance it's hard to resist taming that power. You pat a dog and then smell your hand to smell how much dog is left on your hand and what that dog smells like. It's the most concrete information you'll have all day.

When I asked Coty's Carlos Timiraos how much a successful perfume makes, he said, "It depends on your definition of success. It takes a lot of money to make a fragrance work, and there can also be huge differences between grosses and profits. Industry profitability? Generally the investor's risk is the same as stocks; you make about as much money as you would in the stock market. A good profit for a perfume is generally a return in the teens. And you virtually never make money in the first two years, sometimes not for three or four, although people are launching so many perfumes that a very few are now making a profit in the first year or two. And a lot of perfumes lose money; again, it's very much like

Hollywood. Studios can spend a hundred million producing and marketing a movie and make fifty million at the box office. Luckily a perfume has a longer life span than a movie and translates more easily outside the U.S. The superhits, one of the big ones, could on annual sales of a hundred million make between twenty or forty million in net profits in a year. These are the big established megabrands—it could never be a new brand—but it completely depends on many variables. And even a six-month-old megabrand launch will only make money three years down the road."

The surefire formula for making a bestselling masculine seems simply to be mixing together enough dihydromyrcenol (laundry detergent) with the smell of metal garbage can to choke a horse, then topping that with the scent of cryogenically frozen citrus peel dusted with DDT and a whiff of recyled plastic. *Chrome* is fit, at a 10 percent dilution, for controlling weeds on your lawn. *Aramis* makes a fine garage floor sterilizer. But following a plan of simply pumping out some metallic doesn't always work. All sorts of things that smelled of the effluent of arms manufacturing plants were put on the shelves every year and, for some reason, refused to sell. (The List, which is filled with things that smell like the effluent of arms manufacturing plants, doesn't explain this mystery.)

Then there are the perfume industry's odd and increasingly brutal economic practices, which have been rapidly changing— "deteriorating" is what you more or less universally hear.

The oddness starts with the amount of money the perfume brand has to give the Big Boy to develop their new perfume. The amount of money is: zero. The Big Boy pays for everything. Alexander McQueen and Bulgari and Stella McCartney pay literally nothing for the immense amount of time and very expensive work perfumers and the marketers who support them pour into their submissions. They pay nothing for the dozens if not hundreds of alterations that their creatives demand. (Week one: "Let's do a slightly different gardenia angle." Week four: "No, slightly more different." Week eight: "Maybe not gardenia at all." And then they add, "By the way, we'd like a really long *sillage*," which might necessitate a completely different technical approach to the construction. The exhausted perfumer is thinking: *Now* you tell me.)

This means that the process is cruel to the people who actually make the scents. The Big Boys take all the risk and—at the same time—front all the costs. "The worst of both worlds!" as a French executive observed to me dourly. And now there's core listing, where increasingly the licensees (L'Oréal, P&G, Lauder) force the scent makers to pay them a kickback of, say, half a million just to get themselves on the short list of the suppliers who'll get a shot at doing the next Hilary Duff fragrance. You have to pay to play, and that's just for the privilege of getting onto the field. So they cut costs upstream. IFF used to have six hundred materials suppliers, and they slashed the number by half, which meant that somewhere in Indonesia some small-time patchouli grower saw his market suddenly vanish.

The Big Boys' only growth sector (completely unproductive) is their legal departments as LVMH must rely on Firmenich lawyers and Givaudan chemists to establish, and maintain with constant National Security Agency–level international surveillance, that some bureaucrat in Tokyo hasn't suddenly outlawed Hedione, thus instantly making 90 percent of Dior's perfumes the legal equivalent of poison sarin gas in Japan. (And speaking of the regulatory environment, it is only getting worse. I was traveling around Tunisia with a perfume materials marketer who was talking about the growing List of Banned Materials incredulously. "No more macrocyclics! No more pthalate musks!")

As much as everyone complains, no one changes things. "And how would we?" an American executive said to me. "Our clients, who benefit most from the current system, are exactly the people we can't afford to piss off." Simply because it's the way it's always been done, the industry is structured such that the Big Boys make their money on exactly one transaction and one only: supplying the finished juice. They cover every single cost until that moment, with no guarantee whatsoever of making a dime. It is an exceedingly bizarre structure, as those in it will be the first to tell you, and it has been this way from the start.

This means that the Big Boys are locked into a system of ferocious competition against each other. If you are a BPI or L'Oréal, an LVMH or Estée Lauder, the Big Boys fight for your brief, submitting any number of scents. Which (the Big Boys point out, privately) means a mess: L'Oréal winds up faced with, say, six different perfumers from six different Big Boys submitting maybe five submissions each, and the creatives are frozen before thirty different unfinished proposed scents. Paralysis.

To help them choose, just as the Hollywood studios do with their products, the marketing teams pay a lot of money to marketing consultants to focus-group the scents to death, thirty nice people from a suburb in New Jersey sitting in a room saying, "Nah, I don't like it." It reduces risk. But it usually produces mediocrities. Of course often these are mediocrities like Calvin Klein's *Euphoria*, which haul in money hand over fist till Calvin Klein's licensor is drowning in cash. (The focus groups, of course, cost more money.)

But all of this is nothing compared to the real cost, which is the launch. Now, finally, it is the brands—Coty, for example—that put in the money and assume the risk. Debuting a Coty perfume means Coty does media buys, publicity, samples, and training. One European man, a veteran of numerous perfumes, estimated to me that the minimum euros a house needs to invest in media for any correct launch is €2 million, although I was at a dinner party at which an American woman very powerful in the industry stated that it was more like €18 million or it wouldn't work.

"Media is national," said the veteran, "then the retailer pays for placement in the stores on a co-op basis. You can't launch unless you're in *The New York Times*, which is $70,000 a page, or maybe $125,000 a page depending on where you want your placement. Then there are all the other newspapers and magazines. In Europe we love the *affichage*, ads at bus stops and so on. In New York this is called guerrilla marketing and it's looked down on, but you need it in Europe because the buzz is very important. For Diane von Furstenburg we put her perfume on the tops of all the taxis, and that was fun. In the three top cities in France *affichage* starts at $150,000 for one week and can rocket up from there to three million euros. And TV—forget it. And then there's the GWP [gift with purchase] to budget for, the POS [point of sale] materials to pay for, the visuals, the displays, the posters, which is a huge cost, PLV [*publicité sur le lieu de vente*, the on-sales-site ads], and co-op.

"Co-op is the pretty notorious practice of the retailer paying half the cost of the ads in the store's catalog. Say Hermès does a special co-op with Macy's, an ad with them, in *The New York Times*. Macy's says they're sharing the *Times* ad, except that Macy's has a huge discount with the *Times* because of volume, and Macy's places the ad, not Hermès, so it actually costs them almost nothing. And then they say 'And we'll share the cost of our catalogs,' which, Jesus, are a pure profit center for them, let's be serious, you pay $30,000 for a page ad in Macy's catalog, and it actually just brings them money by using Hermès's name and image to sell Macy's to the public, but Macy's buying office is judged on their ability to sell this advertising space in the catalogs. They start with the top houses, the most prestigious, and go down till all their pages are sold.

"So you launch at Macy's Herald Square, which is the number one surface in the world in terms of perfume sales, they sell you the Seventh Avenue outpost (and you pay for a new carpet to be put down the center of the aisle), and you hire twenty people to demo the perfume—spritz people as they walk by—and you have two or three people, whose entire salary you pay, working at the store to make sure everything goes right, and that's half a million dollars for one week. And if you don't do all this, then you're not a Big Player, and your projections for the year are going to be less. And the sad thing is that the minute your seven days are up, everything you've paid for is ripped out, your new carpet is tossed, and the next fragrance invades the space." He rolled his eyes. He's been through it a thousand times, and every time is a gamble.

To make the retail game more complicated, there is what is known as the parallel market. As a French executive in Paris once expressed it to me, "The biggest, dirtiest secret is that you have to sell your products to yourself." He explained that Chanel France makes the products and sells them to Chanel USA, which does the marketing and sells the products to Neiman Marcus or Saks. "That's what's official, OK?" But Chanel France also sells to Dubai, to a distributor supposed to promote the products in Dubai. And what if Dubai suddenly sells its products to some dirty stores on Fourteenth Street. "They shouldn't," he said with a shrug, "but of course they do. Chanel and certain other brands have trademark restriction, and they mark the boxes 'Dubai' or 'Poland' to identify where they were sold, and when you're Chanel and you find your Dubai-destined perfume in some crummy West Fourteenth Street discounter you go back to Dubai and say 'Fuck you, you're a bunch of dirty Diverters,' and shut them down."

To control diverting, the Chanels of the world used to use lot numbers on the bottles. The Diverters ripped off the shinkwrap, opened the boxes, altered the numbers on the bottles, put the bottles back in the boxes, reshrinkwrapped them, and sent them out. Then the Chanels began writing the numbers in special invisible ink. But the Diverters bought black lights to find the codes and packages and either just cut them out or altered them with the same ink, then sold them on the black market. "And there's new technology all the time," another French executive told me. We were talking about this at Isetan, the huge department store in Tokyo, and one of his principal responsibilities was dealing with the problem in the Asian markets. "Now they also make a genetic thing you put in the juice itself, it floats around and you can read it simply by pointing a gunlike machine at the juice and pressing a button; Gaultier is very good at that, and they can't get the stuff out of the juice, but they'll find some way around it. Chanel is the best at protecting themselves," he said. "LVMH is not especially good. If you take all the sales, 50 percent of perfumes are sold in these parallel markets. It helps Chanel France's bottom line, but it hurts Chanel USA's bottom line. Nowadays it's such a common practice it's become uncontrollable. And even the parallel market is in its worst state everywhere because heads of companies have to make their yearly numbers, and so at the end of the year they get desperate and start pouring product into the legitimate markets, which then overflow into the parallel market, which right now is actually flooded. But at the same time it's a sign of success: What's the point of dealing in these products if no one wants you? One time my boss came in and pumped his arm in the air and shouted, 'Yes! We're on the black market!' "

In fact, the distribution system is now so overloaded that retailers are requiring houses to take back what they haven't sold, the discrepancy between "selling in"—what Hermès might sell to its distributors for placement in stores—and "selling out"—how much Hermès perfume clients actually bought. This is new.

Everyone wants a hit, a *Flower by Kenzo* or a *Pleasures*, but it is an expensive roll of the dice every time you try to launch one. There is one long-time protection that has grown up in response: The perfume industry's accounting rivals Hollywood's in ingenuity, and the LVMHs and Estée Lauders, with their subsidiaries, numerous brands, and vast spreadsheets, could work any numeric magic they wished. When you are a billionaire like Jean-Louis Dumas, Ronald Lauder, Bernard Arnault (head of LVMH), or Alain Wertheimer (owner of Chanel), you are careful to blur or mask any numbers going out to the public that might reveal a flaw. You have a compelling interest in doing so: These are empires resting on nothing but image. (Naturally it is slightly easier to get the good sales figures, but no one trusts those, either.) I have been given numbers by CEOs so ridiculous I would never repeat them. One president gave me figures on his rival, Chanel. Are they accurate? Who knows. The company is privately owned, and Chanel numbers are more masked than any others.

The careful masking is, of course, the way of dealing with the flops. "You can't always tell a flop," a French perfumer said to me, "and no one can find out exactly how much a perfume lost because they consolidate their figures, although," he added, "everyone whispers guesses." The strategy, he said, is that the brand spends in the first year on ads what it expects to gross in that year on the perfume. "There are some that everyone considers clear successes: *Acqua di Giò Pour Homme*, which Jacques Cavalier of Firmenich did. I hate it, but it's genius. Fresh. Clean. Crisp. And it *smells*, a huge diffusion, massive volume. It's a work of art. It's magic. It's up to €150 million total worldwide and was number one on the lists for five years. *Le Male* by Francis Kurkdjian is a huge hit. *Cool Water* by Pierre Bourdon, hit, *Pleasures* by Alberto Morillas and Annie Buzantian, hit, *Polo Blue*"—by Carlos Benaïm, Christophe Laudamiel, Laurent Bruyère, and Pierre Wargnye; perfumes are increasingly resembling Hollywood screenplays-by-committee—"hit, both *Obsessions.*

"Of course, you never know how much ad money they've poured into them. Well, some you just know. Did Lauder spend a hundred-twenty million in advertising for *Beyond Paradise*, or did they spend two hundred? I remember it as two hundred. Chanel's

Egoïste I heard is a flop given the amount they spent on advertising. *Higher for Men* by Dior? No one will give you figures. *Champs Elysées* by Guerlain is one of the paradigmatic, legendary disasters. *Alexander McQueen* is a *huge* flop. Some of them you just have to assume. *Glamourous* by Ralph Lauren? It's not even listed! Christian Lacroix's *C'est La Vie* was a huge flop because they launched, expensively, right after launching the haute couture house and LVMH thought he was going to do so well, and it didn't. *cK Be* tanked in the U.S. Armani's *Giò* by Françoise Caron has a beautiful tuberose note and was a huge flop. *Spellbound* by Estée Lauder was a huge flop, a spicy clove, and *Nu* and *M7* by YSL were disasters."

Guerlain's *Mahora* "was probably the biggest flop I've ever seen," one executive told me. He sighed. "You work like a dog, you get it on the shelf, you spend millions in advertising, and you wind up destroying tons of product in the factory."

Everyone wanted to make the List. It was only natural. The question was how.

⋙◎

Dubrule and Gautier were, naturally, conscious of all the variables. They were also facing a growth rate in perfume sales of about 3 percent a year, which was anemic. It meant (everyone knew this) that any growth would come from pillaging the competition. It was brutal, and getting more brutal to a great degree because the industry had shifted to a new, aggressive, policy of developing fragrances that would appeal to the maximum number of people, which meant fragrances offering no resistance—sweet, coy things that didn't just sleep with you on the first date but in the first ten seconds. The problem was that these scents were like Roman candles, a swift, awesome rise in sale (boosted by all your marketing expenditures), fifteen minutes of fame, and a swift crash as the next one got shoved onto the market. You generated large volume from "global appeal" and sudden death because each of them smelled exactly like the other. Why shouldn't the client jump from launch to launch? Banality.

The second problem from Dubrule and Gautier's perspective was that they were—oh yes—in the *luxury* business. That meant exclusivity. But the industry guru Yves de Chiris had been observing that everyone was acting like mass marketers. The fragrance problem was that the client couldn't "reference" the truly original perfume, and what consumers couldn't reference worried them, and what worried them was not commercial. Consumers claimed they liked original fragrances, but they got out their credit cards for the familiar.

De Chiris had a response, which was Thierry Mugler's *Angel*. The scent was created by the perfumer Olivier Cresp, with de Chiris serving as one of the artistic consultants. The secret of *Angel*'s formula is a juxtaposition of two diabetic-shock-inducing sweet scents, marzipan (created with a synthetic molecule called coumarin) and cotton candy (the molecule ethyl maltol; this is the molecule you're actually tasting when you eat cotton candy). Alone, this would be a tooth-rotting confection, but Cresp threw in a huge dose of natural patchouli, a grasslike plant grown principally in Indonesia, and its strange strong green/organic smell cuts through and freezes the sweetness "like," de Chiris put it nicely, "crushed ice in Grand Marnier."

Angel failed for two years—consumers found it incredibly divisive, pure love/hate. The house refused to drop it. Then it started to build. Then it exploded to become one of the best-selling perfumes of all time, and it is still on the top of the List. (Incidentally, *Angel*'s secret is more or less identical to Coca-Cola's secret: Coke's flavor comes from an overdosed sugar juxtaposed on a superacid taste.)

But the secret of *Angel*'s commercial success, de Chiris argued convincingly, was exactly this divisiveness. Only 3 percent of consumers would touch it. But those 3 percent were complete fanatics, enough to put the thing repeatedly in the top ten. So here was another model to think about: Go narrow and deep, proposed de Chiris, rather than wide and shallow. Which was another way, you could argue, of keeping perfume luxury: innovative, edgy, and not for everyone.

Gautier and Dubrule were aware of the strategy used so successfully with *Angel,* and they were in the midst of their own tactical defining of "Hermès perfume." "When I arrived at Hermès," Gautier told me, "we were going in all sorts of directions and were not as strong and coherent as we could be. So I took us in a direction completely different from the rest of perfumery and the rest of marketing." Gautier was given to repeating, with the conviction of a mantra, that the error the luxury houses made was to do marketing for *grande consommation,* which essentially consisted of analyzing a demand and then creating a product that, one hoped, anticipated it. She was not that. She was Hermès, and Hermès was luxury, and luxury's logic turned that theory upside down by starting with *l'offre,* an offer that would create a desire, and a demand. "It's the opposite of marketing," stated Gautier simply, with the assurance of an economics professor.

Dubrule's comment was that in true luxury, one worked with artists, creators on whom one placed few constraints. One worked from an artistic vision, not (she said it with only slightly less disdain than did Gautier) from market studies. Which she had done; she had a business degree from one of the best business schools in France, thank you, and she knew perfectly well how the focus groups and so on worked. But to her—and to Gautier, who stated it adamantly—luxury meant an if-you-build-it-they-will-come strategy. Luxury, to Gautier, was *proposé.*

This was the subject of the meeting I sat in on.

We sat at the round table in Dubrule's office, a long, clean, practical workroom with her desk at the far end by the window and the round meeting table and whiteboard at the other. They talked, and I took notes on my computer. Dubrule slashed arrows and arced big circles of emphasis around words on tablets of paper that represented the various Hermès perfumes. By prior agreement between myself and *The New Yorker,* on the one hand, and Hermès, on the other—we'd established some logical ground rules regarding the disposition of industrial proprietary information and Hermès trade secrets I would learn in my reporting—I can write about none of the details, but the general question before them was, when they considered all the above variables, How were they to place Hermès perfumes in the global market? The overall thrust, which Dubrule laid out, was that she and Gautier had decided on a three-pronged approach.

On the highest level would be the *Hermèssences* collection, superexpensive, superluxurious fragrances. The *Hermèssences* were scheduled to launch in October 2004. Ellena was hard at work on them. The house was going to start the collection with four scents—*Rose Ikebana, Vetiver Tonka, Poivre Samarcande,* and *Ambre Narguilé*—and add to it, one by one, every few years. Dubrule described their concept as being like the experience of dining with Pierre Gagnaire, Guy Savoy, Marc Verra, "great French chefs who are going to search out unexpected contrasts. We will," she added with emphasis, "be able to use some very Hermès materials." By this she meant expensive; the fragrances would thus be much more expensive than all the others on the market. The culinary description she meant as metaphor, though Ellena was creating one called *Ambre Narguilé* that would, when he was finished with it, smell— gorgeously, exquisitely—of cool, freshly sliced apples wrapped in leaves of blond tobacco drizzled with caramel, cinnamon, banana, and rum. This collection would only ever be sold in Hermès stores; these Hermès products would, significantly, never be *sortis de leur univers*—taken out of Hermès's world. Below this price point would be the fundamental collection, *Calèche, Rouge,* and *Hiris;* the *Jardins* collection would be at a very slightly lower price point than *Calèche* and equal to that of *Eau des Merveilles.* Gautier was talking darkly about Hermès's prices in the United States. She would have to raise them about 15 to 20 percent to compensate for the decaying dollar.

And that was the plan.

CHAPTER 7

ON NOVEMBER 22, 2005, when serious winter has already set in on New York City, bringing battleship gray clouds, intermittent cold rain, and a chilled wind off the metallic Hudson, the first creative meeting on Sarah Jessica Parker's perfume product is held at IFF's global headquarters, 521 West Fifty-seventh Street.

It is 3:00 P.M., and you'd think it was 6:00, or maybe 8:00. At mid-afternoon the cars already have their headlights on. The headquarters, which sits opposite the New York CBS affiliate, is a paradigmatic massive nondescript office structure with 1980s gold-colored doors. Donald Trump meets East Berlin. Catherine Walsh and Carlos Timiraos are at eighth-floor reception. Several IFF people are gathering as well because Parker is arriving. (One says to me discreetly, "It's not my project, but I really like *Sex and the City*.")

Everyone goes downstairs to wait for her. Joanne H. Trembley,

IFF's VP of sales for North America, is there, and Yvette Ross, IFF's senior account executive on Parker's perfume. Walsh chats with Trembley and Ross and keeps an eye on West Fifty-seventh Street.

Exactly on time, a black town car pulls to the curb and stops, and we watch someone inside open the large rear door. Melinda Relyea, Parker's assistant, a young blond woman preternaturally laid-back and cool, gets out, followed by Parker. It's always a bit of a surprise how thin and small she is. She is wearing stilettos and a trim gray camel's-hair coat with a belt tied precisely around it and, in the afternoon gloom, gigantic jet-black Ralph Lauren sunglasses that cover her face. We watch the chilly wind blowing her hair around. "Why is it so cold so early," murmurs Yvette Ross to no one. Parker comes out of the gray into the warm lobby, and we all start the greetings. She takes off her coat, and she's dressed in a simple blue shirt and slacks. Some mascara and a bit of natural lip color. Her eyes are the same intense storm-cloud blue.

I sort of hang back a little, and when she gets done with the rest, she turns to me and for an instant we both hesitate there in the lobby. I realize she's not certain what is appropriate either, but, journalist subject divide or not, I give her a kiss on the cheek. She beams and says intensely, "How *are* you?" with a voice that has a raspy-squeaky angle. We all crowd into the elevator, and someone hits the button.

The meeting takes place in one of the larger meeting rooms I've seen in the IFF complex, a glass-enclosed room surrounded by the offices of perfumers and evaluators. There's a vast expanse of table, and we do that thing where everyone enters and chooses their places carefully. Ross, as the IFF account exec, directs Parker to the center. ("Here?" says Parker, "you want me here?") I find the chair nearest an electrical outlet, plug in my computer, sit down next to Walsh.

There are three sets of players here, three interdependent king-
doms.

Coty is arguably first among equals. It is Coty, in the person of
Walsh, that allows all this to happen. The financial risk is Coty's,
and they are capitalizing the brand with millions of dollars in development, overhead, marketing, advertising. IFF is quite aware
that Coty is its client. Everyone knows that Coty had sent the brief
to other scent makers, and that Walsh chose the juice created by the
IFF perfumers Laurent Le Guernec and Clément Gavarry—both
of them are here as well—which is why IFF won the brief, and in
this plush conference room the IFF people, in all the appropriate,
subtle ways of corporate interactions, make clear to Walsh that they
appreciate her business. Le Guernec is sitting next to Parker at the
table, telling her about a scent angle he's trying for another perfume project. Gavarry is next to him, smiling broadly.

Parker is the concept, the commercial idea around which this
particular Coty venture is built. Walsh is there because Parker is
there, and it is Parker's aesthetic that's guiding all of this, and her
artistic vision of the juice is what will be sold. She's central to the
commercial idea, but then Parker's only in the room because Walsh
said yes to her. And everyone is aware of the various roles that are
being played and of who ultimately signs the checks. A few months
before, Parker, with Matthew Broderick, spotted Walsh at a glamorous black-tie event at the Plaza hotel. She grabbed Walsh by the
arm and, somewhat to Walsh's consternation, towed her around the
room introducing her to everyone she could corral as "my boss."

"Don't forget the sushi," says Ross. Parker looks around, exclaims, "There it is!" There's a table overflowing with cookies,
drinks, salad, cheeses, fresh fruit (New York delis do great fresh
fruit). Some of us take plates, pick up tongs. Someone sets down a
large bowl of popcorn. I put two pieces of maguro on my plate.
"And it's the *good* stuff," says Timiraos, who is right behind me with

a plate. (It is, indeed, the expensive stuff. We try to look like we're not wolfing it down.)

We reassemble at the table. People are chopsticking up pieces of raw clam and rice, the IFF contingent is organizing notes, and Walsh and Parker avidly discuss Parker's recent appearance on Oprah's once-a-year Favorite Things show. "It's the things that get her through the day," Parker says, "from the necessary to the completely indulgent, soaps to slippers to TiVo. Last year," she adds, the audience "was all teachers."

"This year it was all Katrina workers," says Walsh.

"And she picks one of her favorite things and gives it to everyone in the audience," says Parker. "And she picked *Lovely!*" The big deal with Oprah is if she'll say the name of the product, and Parker tells us about her best friend, Jill, who called her and said (she puts on a thick New York accent), "Dijya *see* it?! Dijya *see* it?! If she said it once, she said it *foive toimes!*"

"She asked me afterward," says Walsh to the room, of Parker, "how it had gone, and I told her that if Oprah had simply said Sarah Jessica had a new fragrance, I'd have been happy. The fact that Oprah, who doesn't wear fragrance, said this is her favorite scent, and that she *wears* it, is amazing!"

This happened days ago, and they're still elated.

It's Parker who claps the meeting to order. "OK! What are we doing today?"

There's a three-part agenda. First, a "what's this new product going to be about" discussion combined with a business update on the state of the brand. Second, a crash course for Parker in cosmetic chemistry; after years in the industry Walsh and Timiraos generally know stearic acid from propylparaben, but Parker does not, and as they're going to be creating a perfume product (they'll be determining what kind) in a rather unusual base, they want her to be comfortable with propylparaben as well. Third is giving Parker a little course in Perfumery 101.

Walsh sits forward, and the room turns to her. "I think it's important to go over what we were trying to do when we created the Sarah Jessica Parker brand," she begins. "That this scent would not be created like other celebrity fragrances. We [at Coty] were the masters, the starters of the celebrity fragrances." She looks in turn at each person. "We learned that with other celebrities you have to launch something every three months. But here we wanted something that was the opposite of what people had seen in *Sex and the City*." Parker is watching Walsh attentively. "Sarah Jessica was provocative in that series. Here with her fragrance, it's different. We wanted her perfume to possess a sense of quiet. An understated elegance."

Parker says, "A trust in the audience. That assumes, as good television does, that they're smart."

Ross and Le Guernec are taking notes.

Walsh puts the first big point on the table. "What we're doing today is not"—she stresses the *not*—"coming back next fall with a Sarah Jessica Parker flanker."

Flanker is the industry term for a new version of an existing perfume. *Polo* is a hit; Ralph Lauren puts out *Polo Blue*. Privately the perfume industry views flankers the way the movie industry views sequels; like later *Star Wars* episodes, they enjoy rather lesser reputations. Financially, however, they have a logic, though sometimes the flanker happens because the original was a success, sometimes because it was a failure: Chanel did *Egoïste*, which disappointed, but having invested so heavily in the name, it did *Egoïste Platinum*, which succeeded. Often flankers are dictated by the investment in the name. Yves Saint Laurent poured a river of money into launching *M7* (the name stands for "the seventh YSL masculine"), created by the star perfumers Alberto Morillas and Jacques Cavalier of Firmenich. *M7* smells like a Fiat engine engulfed in flame on a shoulder of the A6, an alarming chemical storm of burned rubber, charred metal, torched leather, and toxic melting polycarbon. This is not necessarily a criticism: It was a well constructed, thoughtfully built scorched car in flames. But people stayed away by the million, and the scent was a disaster. YSL allowed some time to go by, then launched the *M7* flanker, *M7 Fresh*. *M7 Fresh*, also by Morillas and Cavallier, is an excellently constructed light smoke scent, relieved of the flame-thrower death smell, mesmerizingly delightful and in all ways better than the original. (It too failed to sell, however, and Yves Saint Laurent has now, in the best Argentine manner, quietly disappeared both scents.)

Flankers are supposed to be variations on a common theme, but the only thing *Eternity* and its flankers have in common are the ad dollars Calvin Klein marketers spent on imbedding that name in people's neurons. Dior is king of the flankers with its *Poison* franchise. *Poison* of 1985 was followed by *Tendre Poison, Hypnotique Poison, Pure Poison*, and, in 2007, *Midnight Poison*. I once asked a Dior exec, What is the commonality between the *Poisons*? "None," he said. "None! Well"— he reconsidered it—"all have white flowers," and then he listed the flowers, which bore no similarities. I said, In other words, none.

He said, "Correct."

Flankers differ from seasonals in that flankers are permanent (that is, they're allowed to live as long as they sell) additions to the house's collection, whereas seasonals have death sentences built into their designated life spans. (Seasonals are mostly issued for summer and are basically excuses for the house to bring out lighter, more commercial perfumes.) Escada is the king of seasonals. Puig, Escada's licensee, terminated *Ibiza Hippie*, a summer seasonal with a flanker-sounding name. It was a terrific little commercial jewel, and deleting it was a minor crime.

Walsh wants Sarah Jessica Parker's *Lovely* to be what Parker wants it to be: a classic, a serious, elegant perfume for adults. The dilemma is that they also want to put out a new product that reincarnates *Lovely*, this perfume they've worked so hard on, and to the market, that says flanker, but Walsh is determined somehow to produce a sequel that is an original at the same time. She is saying to them that they are going to walk this tightrope, and they're going to walk it successfully. The question is how. "We want an initiative other than a flanker," says Walsh. She emphasizes the words, looking at each of them, and the table is rapt. "This product we start working on today is not about changing *Lovely* in any way. This is about using *Lovely* to give the customer something different. It will have the same olfactive heart of *Lovely*, but it's going to have to be big enough and strong enough to stand entirely on its own. We have nice ancillaries in the line. This has to be perceived by the customer as just as important as the first launch [of the perfume], yet keeping the fundamentals of the scent."

She pauses. "Remember that Sarah Jessica created her scent from scratch, from components in her head, with a very particular vision, a scent that smelled of skin, and there's something in your concoction—and I use that word as lovingly as possible—of those three elements that is far from what we put in that bottle. That something is what we're going to find today. That something is our product." That, in short, is why this thing they're starting right now will stand on its own. The same perfume, but with a different aspect of its soul made visible.

She turns to Parker. "Remember how you said it was too dirty, so we flowered it up, and you said it was too flowery."

They both laugh.

Parker: "Too chaste."

Walsh: "Too fat."

Parker: "Oh, god!"

Walsh: "And we got to something that is so clearly you and so clearly your vision but that simply worked better as your first scent. Better for your audience."

Parker nods.

Walsh: "But still there's something you're not getting out of this scent."

"Something," Parker turns to me and carefully stipulates, "that we agreed not to get. Knowingly not getting it. There was a certain fattiness that brought down the higher notes and a bit of dirtiness. And honestly"—she turns to them now—"I don't think that would have been right."

Walsh says to me: "Think of it this way, Chandler. The component of Sarah Jessica's original concept that carried the most character was the African oil, which brought the fattiness, the human skin; the other two things brought smoke-plus-dirt and, then, the girl." To Parker: "When you lose that oil, you just lose that skin aspect."

Parker: "The perfume just gets higher!"

Timiraos says to the room, "We felt that African oil was the best starting point for this reinterpretation of her fragrance."

"So let's not take our eye off *Lovely*," says Walsh to everyone. "Let's return to our original story but create something technologically that went missing with that original oil. Because it's Sarah Jessica, we're going to be launching at the busiest time of the year, but this project can't be 'Here's a nice little gesture.' It has to be a big idea. Not an ancillary or a soft launch. We are totally trying to do something that the market is not doing. We presented this to Michele [Scannavini, president worldwide of Coty Prestige] and Bernd [Beetz, CEO of Coty, Inc.], and they're totally behind it."

Walsh pauses. Everyone is listening intently. "Great," says Parker, focused on Walsh.

⤴

They formally start the business report on *Lovely*.

"We picked one brand we want to make our 'classic,'" says Walsh with an eye on Timiraos, "and it's this brand. Now, we did something very smart with the launch in the market phasing-in." She means the delicate and complex strategy of choosing media markets and placing ads carefully in specific retail spaces to benefit the product. "Most people come out with everything immediately, all the ads, print, broadcast, editorial. Launch hysteria. The perfume goes to number one, and then there's no more marketing support behind it, and you watch it peter out. That's exactly what we didn't want to do with Sarah Jessica. We had a very interesting phasing to *Lovely*. We launched in September 2005. Now it's late November, and it's almost like we're still launching."

Parker says to me, "We haven't even launched the on-air yet." I say, Really? I'm pretty surprised; I have the distinct impression of having seen the broadcast ad. She shakes her head, says, "Yeah, those of us who live in New York think we've seen it"—the *Lovely* ad images were, for example, two stories high on Lord & Taylor on Fifth Avenue—"but we haven't."

Timiraos: "We didn't play all our cards up front." He grins. "I have to say, there were moments in the beginning when we were starting to doubt ourselves." Parker laughs. "I never told you that," he says to her. "I remember some calls to Catherine, 'Did we do the right thing pacing the launch like this?'"

Parker says to Walsh, laughing, "All on those tiny narrow shoulders!"

Walsh shrugs her narrow shoulders. "You have to stand behind your work. The specialty stores were no problem, but we got a significant amount of resistance from the department stores."

Walsh also has to manage international markets, and with Parker's perfume, she tells me later, she'd had to do some smart managing "because Asia [that is, Coty's various Asian distributors] came to us and said, 'We'd really like to launch this perfume, but it might not suit Asians so we want to do a gentle launch.' Carlos and I said, 'Thanks, but in that case you're not getting the product. You launch it, or no.' As of June '06 we're in Singapore and Hong Kong and doing very well, but we're not in Japan." (Coty has business partners in each market around the world, affiliates and business units, and the partners make the decision on how to invest and what products to support. Their money can come from them and/or from Coty, the parent, but each country is independent.)

They talk about the skin aspect of the scent. Yvette Ross, IFF's senior account executive, says, "I tell you, Sarah Jessica, there are people here used for skin, people we reserve to test the smell of perfumes on skin—"

Parker: "That's so amazing!"

"—and your perfume worked beautifully on them."

Parker: "That's great to hear. It's not unlike what I do for a living. Whatever is formulaic is always what gets copied. It's the remake, it's the endless imitation of, because that's what people feel safe with. If a department store wants that, I understand. So the thing that makes us doubt—the risk at doing something novel, at not being formulaic—that's exactly what we really want to do."

"Which is why we made the marketing for *Lovely* different," says Carlos Timiraos. He explains that Belinda Arnold spent months with Ina Treciokas, SJP's press rep, figuring out exactly "where they'd be," as they phrased it (they meant what stores their visuals would appear in, in which publications their ads would appear, their overall presence) and how to get themselves into the places they wanted to be. "We held back on some PR at first. We paced ourselves." Gradually they opened up a huge and very expensive campaign, but Coty carefully chose their media partners. "We had an exclusive with *Vogue*, that was great, and people are still talking about it, then *Us Weekly*." He realizes, asks Parker, "Have you seen the letters to the editor?"

Parker: "No!"

Timiraos: "I'll show them to you."

Walsh gets up and presents Parker a bestseller chart of numbers from NPD, the perfume industry's *Billboard*. "We're in stage five [of the launch], and we still have several phases to go. Nordstrom really appreciated the scent." (Due to its prestige, Walsh gave Nordstrom a "soft launch" exclusive on *Lovely* for the month of July 2005.) "But Nordstrom doesn't do a lot of theater." She means that unlike Macy's and Bloomingdale's, Nordstrom doesn't do large visual displays. "Now," says Walsh, and Parker's BlackBerry goes off, her ring a series of gentle bells. Parker grabs it and says, "Sorry!" hits the buttons, and the ringing stops. Walsh points at the monthly figures. "Now these are from the fourth week in July till the third week in October." (*Lovely*'s fiscal year as a brand coincides with Coty's fiscal year, which starts July 1.) She points at the figures as she goes; Parker is absorbed. "We were $550,000 [sales] per week here, then $625,000. Then we dipped and you did personal appearances, and see where it went? Then we go up to an $800,000 week when we go on *Oprah*."

Parker asks about a few details. She looks at once encouraged and unnerved, calm but wondering, it seems, where the bad news is. "The figures are great," says Walsh reassuringly, reading her. Walsh is solid, professional.

"They're great," Timiraos says to Parker from the other side of the table. (She looks toward him, back to the figures.) "It's not about launching big, it's about staying there, and we've stayed top ten, in *this* environment, through launches from JLo and Ralph Lauren and Calvin Klein, and we're still there, and it's amazing." He grins, but says seriously, "And we worry about every sale."

Parker faces Timiraos and says, urgently, anxiously, "And I worry about every sale! God, I worry!" She leans across the table as she says it.

Timiraos: "We've taken our net sales plan [the forecasts of what they'll sell based on sales trends] up almost fifty percent!"

This is the result of a huge amount of planning worked out and bets placed. Walsh and Timiraos spend a significant portion of their time and talents calculating marketing variables, generally conferring with Parker only on the most complete issues of how to place the brand. They decide when to ship, how many units. They started with a projection that the Sarah Jessica Parker perfume brand would be a $30-million brand in Coty's annual sales at wholesale prices. (Retailers generally do a 40 percent markup; that is, if Coty wholesales $30 million in bottles of Parker's perfume to Saks, Saks will sell them for $42 million.)

Then they started pouring in money. "As the licensee, to make a perfume you actually build machinery, you build tools and molds, which costs hundreds of thousands of dollars," Timiraos would say to me separately in his office. "On a short lead time, which we had— once we signed the deal, we got her scent to market very fast—you sometimes have to purchase perfumery raw materials before the testing on them is complete, so if they don't pass the toxicity and R&D testing—and thankfully all of ours did—that's a huge risk you take. You'd have to redraw your formula. We really moved at light speed. And there's a huge opportunity cost; if you choose the wrong celebrity, you'll burn millions of dollars for no return, so you have to be very careful. The launch will cost you two to three million dollars at a minimum, then annual advertising and marketing costs tens of millions. The first few years you're just putting in millions and millions." He stopped and sort of rolled his eyes and sort of smiled. "You know," he said, looking directly at me, "in the end? It's all a big crapshoot. Do all the research you want, until your opening weekend grosses? None of it matters."

In the IFF meeting, Walsh reminds everyone that they started with an estimate of $30 million in sales annually. They soon discovered this was wrong. "Quarterly," Walsh says, "Carlos and I sit down and say, 'OK, what's the status of business today? How are our markets?' and then revise our forecast. By the first quarterly review we already had to revise upward our projection for the end of the year. It has now gone up to $46 million." (Four months later, Timiraos will tell me that it had grown again to well over a $60-million gross.)

"Wonderful," says Ross, congratulating Parker.

Walsh: "And we haven't gotten through Christmas." Everyone is smiling. Walsh and Timiraos are happy, but Parker seems mostly relieved, palpably so.

Change of gears. Walsh now shows Parker the NPDs for the first time. The List of the top hundred best-selling perfumes on the market by week, month, and year. Parker is fascinated. Walsh tells her that Hess, her agent, has been getting the NPDs and, ever protective, first gives her some caveats: The List is manipulated in all sorts of ways and for all sorts of reasons. For example, spiffing. ("Spiffing," says Parker, looking at the columns of perfumes and the columns of rankings and of retail sales.) Walsh explains "spiffing," a marketing strategy that is, in essence, payola. It goes like this. The store manager tells the BAs (the beauty associates, the sellers that Coty in this case pays per hour to stand on the floor of Macy's and push the product) that today, they'll get another five dollars for every bottle they sell. More bottles are sold, so volume increases, but spiffing takes money away from Coty—and Parker, if she's being paid in a percentage of sales—with each sale.

And the List accounts for less than all sales. No one can be sure, because of the way data is collected, exactly how much of the total it represents. (At the time of this meeting, Nordstrom and Nieman Marcus, for example, didn't report to NPD.) "Still," says Walsh, "this is 1800 doors." (A "door" is a point of sale, or store.) She points at a column. "This is the month of October." She indicates gray boxes. "What's in gray is what's sold at the fragrance bar, meaning you have your six feet of space at Bloomingdale's, Donna Karan has her six feet." (She looks up to stipulate, "We do not own the person behind the bar." That person is the employee of Bloomingdale's. She looks back at the figures.)

"The white areas here are what the industry classes as the cosmetic brands—the Cliniques, the Lancômes. We don't use those numbers for comparison purposes, for making our List. We want to compare apples with apples, so we remove the cosmetics numbers and only use sales at the fragrance bar." (Coty is not alone in doing this; most of the licensors pick and choose their comparison methods, which is, unofficially of course, done strictly to make their numbers look better.)

"OK," says Walsh, "so here we are. Number one is Calvin Klein's *Euphoria*. Number two is *Be Delicious* by Donna Karan."

Parker: "Oh, that's Trey Laird!" Laird had also done the Donna Karan design and campaign.

"Number three," says Walsh, "is Britney. Number four is *Chanel No. 5*. And the top-five perfume is . . . you!"

Parker looks elated for an instant, then suddenly sardonic. She says with a tough New York accent, "*Chanel 5*, kickin' my ass."

Walsh: "You're crazy!"

Parker rolls her eyes. "I'm kidding, I'm *kidding!*"

Walsh: "No, really, look at the difference between you and *No. 5*, it's $200,000." Virtually nothing.

Parker: "Wow. OK. Cool! Wait, who's nipping at my heels?"

Walsh: "Clinique *Aromatics Elixir.*"

Parker shrugs, very macho: "Eeeeh, no problem." She eyes the list, going down it. Grimly: "Hm, Paris Hilton."

Timiraos: "This is where the spiffs really make a difference 'cause if Donna Karan is spiffing, the girl's gonna sell Donna Karan."

There's a sound at the glass door. Steve Semoff, IFF's vice president for technical services, North America, puts his head into the room. Walsh looks at her watch; he's right on time. "Five minutes," she says. Part 1 is finished; they'll start Part 2 after the restroom-and-sushi break. People get up, stretch. Investigate the popcorn bowl. Walsh and Parker catch up a bit.

✧

One time I saw Parker, she'd just gotten back from Los Angeles filming *The Family Stone*. It was the first studio feature by a young director named Thomas Bezucha, and when I asked her about him she said emphatically, "He's the best director I've ever worked with, and I'd do anything he asked me to." I found it so startlingly categoric, given all the other directors she's worked with, that I asked her if that remark was on the record. She assured me it was.

I went to see the movie at the AMC Empire 25 on Forty-second Street and Eighth Avenue, then called Bezucha's agent, who, like Parker's, was at CAA. Bezucha, when he called me a few days later from LA, turned out to be startlingly nice and low-key. He talked about the way people in Parker's position live in "another reality," he said, "in which every element revolves around that one single point, and I always find with Sarah Jessica that the way she spins her universe, she always works double time to make *you* the center. I find her very deflective of attention. There is a very, very well-mannered and precisely polite personality. But she's also the person you want to be sitting next to in the bleachers at the pep rally, talking about the other students. There was a girl in my seventh-grade French class, Lucy Luddy, and we always had to be separated. I feel that way about her.

"My entry into her world was unusual in that I don't have a tele-vision, and I'd never seen *Sex and the City* when I met her. I was a fan from film. She and I met when I was in New York casting one of the first incarnations of *The Family Stone.* They'd stationed me at the bar of the Soho Grand, and I felt like I was hosting my own TV show because it was a different star every thirty minutes. She was my last meeting, and it was 5:30 P.M.; this was the week before Christmas, the first Christmas after 9/11. The bar had been empty all day, and suddenly—I told her this story once, in the back of the car on the way to the movie's premiere—it was full of people after work. She walks in. In New York, the celebrity thing is funny: You're never, ever supposed to notice, and nobody turned around, but everybody froze. She came over and shook my hand, introduced herself, and it was really interesting to me—there'd been a lot of celebrities, and it wasn't, actually, that she was a famous actress; it was like the essence of all of New York's best qualities, its style, its wit, its elegance, was in the room. It didn't feel as much like people wanted to take her picture as that they wanted to salute."

He talked about working with her, which was, he said, "so easy.

This is only the second film I've done. She's very offhand about her work. 'Oh, I just do what I do.' Which is so obscuring of what she is. You make the smallest suggestion, and she will make that adjustment, and she'll do it, perfectly, till you give another direction. She's a calibrated technician, in the very best sense. She was so protective of that script and of me. Luke [Wilson] is so—he just feels his way, freewheeling. She was precise, completely prepared. The boys had ad-libbed a few things. When Craig [T. Nelson] says sexual orientation is like handedness, Luke ad-libbed his reaction, and in the charade scene, Dermot [Mulroney] and Luke ad-libbed with each other. And then, in the scene where Sarah Jessica is trying to pick mushrooms out of the dish she'd made? And the viewer knows that Dermot's character is allergic to mushrooms? That scene was just supposed to end with the line 'Are those mushrooms?' And Luke says the line, and suddenly she screamed, 'I didn't know!' You can actually see Luke laughing in the shot. That's when I fell flat in love with her. She'd finally felt comfortable enough to create on her own terms, and when she did, it was what made the scene.

"The dark hair was one of her conscious things. She really didn't want to be Carrie [her character from *Sex and the City*]. I'd assumed she'd be blond. She decided to be darker. People assume we talked about it. We didn't. She created her character. I did want her hair up, and I wanted the movie to open on the knot on the back of her head, and she'd be in a gray Kim Novak suit. We sat down to talk about the opening, and she said, 'OK, I'm thinking of wearing my hair up,' and I'm like, 'Uh huh.' And she says, 'And a gray suit?' And I'm like, 'Uh *huh.*' And that was it.

"She had to deliver a line once, in a car, to Luke, a single word: *what.* She did it, and I told her, 'OK, here's what you're feeling. Now put that in.' And the next take, the *what* had totally transformed." He paused to remember the word, the way she'd said it. "It was perfect," he said.

People arrive back precisely on time.

They're in the second phase of the meeting, the beginning of the discussion of the new product: What version of *Lovely* will they make and how will they make it. Development starts now.

"OK," says Ross when everyone is sitting, "this is what Steve calls Cosmetic Chemistry 101." On cue, Semoff enters. He is a very tall man, massive, with thick, dark hair, the slow, deliberate movements of a professor, and an ironic, bone-dry wit, all of which gives him the demeanor of a sardonic, slightly sleepy Jewish grizzly bear. He deals with the technical aspects of products IFF manufactures and knows butylene glycol from sodium laureth sulfosuccinate and so is here to teach Parker about various possibilities for the new *Lovely* product—what gels, silicones, or oils, what serums they might use, the perfumable bases available on the market.

From the various materials Semoff will show Parker, she, Walsh, and Timiraos will choose the ones they want for the new *Lovely*-scented product. Semoff sits next to Parker at the large oval table, and she scrunches over for him.

"So this," says Ross, basically meaning Semoff, "this is, 'Now that we're making your next one, how do we do it?' What does 'emolliency' mean and so on." Semoff blinks sleepily. Mr. Emolliency.

Parker says to everyone, "So that I can be clear: This is about skin, skin, skin. Yes?" She means their working idea for the product: something that will work particularly well on, or render silken, or beautify (they're not sure what it will be yet) skin.

Walsh nods. "This is the first meeting so we can have the same vocabulary."

"You're a consumer," Semoff says to Parker, eyeing her skeptically as if he wasn't quite sure it was true. "So you know what you like. Fine. But the thing we're talking about *here* is how do you communicate that to the guys in the white coats." He pauses, adds dourly, "So they can tell you you can't have it."

Parker, daintily: "Ah. Great. I look forward to that."

Semoff starts his PowerPoint presentation. "This product of yours. You've got five choices." Everyone looks at the screen. "You can only have a liquid, a gel—like a shampoo—a solid—like a candle—a semisolid—say, a very heavy cream—or a powder. And you will have to marry it to packaging: Again, you have a choice of a bottle or a jar, a squeeze package, a pump, a spray, an aerosol, a dropper, a stick, a roll-on, or a wipe."

Parker asks him about the difference between a spray and an aerosol. Semoff explains that a spray is a natural pump, no propellant, the fragrance atomized through the pump mechanism, whereas an aerosol uses a propellant, a liquified natural gas under pressure, usually a hydrocarbon blend of isobutane and propane. The industry uses a fragrance grade that's odorless and tasteless. She nods fast; "Got it, got it." Moving on.

Walsh: "So we've got to choose from this."

Appearance, Form, and Format, says Semoff, and you marry the three. For example (he glances at Parker to make sure she's with him), a clear liquid in a pump spray. The appearance is "clear," the form is "liquid," and format is "pump spray"—basically this is about packaging. You want to create a serum scented with *Lovely*? You might choose translucent plus gel plus dropper. Or you can have a pearlescent powder in a stick.

There are several bases they could do:

>Lipophilic base (oil system);
>Hydrophilic base (water-based system);
>HLB blend (an emulsion system that blends the two).

Semoff then explains that *emulsion* means two insoluble phases residing side by side on a molecular level, one phase continuous, the other phase discontinuous. Oil mixed into water (O/W) or water mixed into oil (W/O). O/W is water is a continuous phase, says Semoff, water as the primary player, and those are lighter. Noxzema Cold Cream is W/O, water mixed into oil.

"Noxzema!" says Parker, grinning. "Crazy!"

Ross prompts Semoff: "Salad dressing."

Oil and vinegar, says Semoff. Two incompatible materials that are mixed on a molecular level will separate. The oil will be suspended in water, but only for a few moments. *Emulsion* simply means a permanent suspension. It's easy, you just put in waxes, or perhaps fatty acids. "Add the tiniest bit of mustard," says Semoff, "and the oil and vinegar will be emulsified; they'll mix permanently. Mustard is an emulsifier." He pauses for a moment of thought. "Probably the starch in it," he murmurs. "Most skin-care products," he says, "are delivered as an emulsion, an oil in water brought together in a single product. Like salad dressing."

Next PowerPoint. Basic Cosmetic Definitions. Semoff begins with emolliency: Something that applies slip to the skin, can be oil or water based. *Dry feel.* Gloss. Spreading. Substantivity. He reaches around and out of the air brings out a bottle of Lubriderm, slaps it down in front of Parker. "Classic greasy." He puts a bottle of Clinique M Lotion on the table: Dry feel. I comment that I really like M Lotion, it's been around for years, and it's still just about the best thing you can buy, effective, no grease, works fast, even during New York City winters that just suck the moisture out of your skin. Parker replies, in good didactic form, "What's the best way to moisturize your hands? Soak them in water for half an hour, then—without drying them—lather on Vaseline and wrap them in Saran Wrap." She adds with a loud whisper, "And that's *free tap water.*" Semoff nods rabbinically: This is truth. Sarah Jessica jumps up to get a cookie, turns back to whisper to me, "Don't tell anyone!" A few people take advantage of her lead and stock up. She adds fruit and popcorn to the cookie. People sit back down.

Semoff starts the next phase: Key Cosmetic Components.

"The key ingredient in everything is water, usually about seventy-five percent. There's a little water in *Lovely*—"

Parker: "In the perfume."

Semoff: "—in the perfume, because it controls the volatility. If it were pure alcohol, it would flash right off the skin and take some of the fragrance with it. It's also in there for accordance with the VOC—Volatile Organic Compound laws of California—laws about the max percentage of volatile products."

Semoff tells Parker about humectants, hydrophilic materials that attract water. He selects glycerine and propylene glycol. "They're put in for feel and to attract moisture. If you're in Tucson, you don't want too much of these because they're moisture magnets, and if the humidity in the air isn't high enough they can also pull moisture out of your skin. In Atlanta in July you have no problem."

There is a game being played here that is kept tacit, though every player, except Parker—they have no need to bother her with these sorts of details—knows it's going on. IFF makes its money by manufacturing *Lovely* and selling it to Coty. Parker's new product is going to be imbedding the scent in a base much more complex—and much more expensive—than the simple alcohol in which perfume is normally served up. IFF's "core competency" is fragrance creation and simple liquid blending and Coty has its own quite competent makeup and skin-care formulators, so IFF won't be making the base. Although Semoff is IFF, and though he's showing Parker these bases made with IFF products, IFF knows that Coty will be creating the base. And Coty knows that Coty will be creating the base. In fact Coty quietly told IFF before the meeting, "Don't show Sarah Jessica anything that we couldn't make or that's outside the Coty competencies." "Because," as Semoff will explain in his office a few weeks later in his usual straightforward style, "then Coty'd have to buy it from us. And that ain't gonna happen."

The creation of bases is a fine science. It's possible to create a shower gel that would "enhance fragrance performance"—show off the scent, in plain language. It will foam and clean, too, but the problem is the surfactants. You need to add surfactants to the base—they give you foam, a nice skin feel, and mildness; they're called betaines—but betaines can cause problems if you're not careful: If your base is alkaline, they'll release free amines (compounds with nitrogen in them) that will smell like fish. (Nitrogens in amine form smell fishy.) The fish smell flashes off quickly, but it's going to wreak havoc with your fragrance's top notes. No one wants jasmine + flounder. "One thing Coty R&D does well," says Semoff, "is create a base that has almost no odor, smells clean. This is why it takes six to eight months to create a product that's ready for prime time."

Semoff talks about "film ingredients," molecules that create a film to allow the wearer to remember the product is there. "Specialty ingredients and fragrance," he says, "these are really what the product is all about. So"—he starts reaching for vials of molecules and gels in lines in front of him—"we're going to look at some things." He assesses Parker. "You got skin?" She grins back at him, nods. "Let's play." Parker moves her chair over to his and happily holds out her arms, wrists up, as if for a phlebotomist.

Semoff starts by showing her a product similar to Chanel Elixir, except that IFF has put *Lovely* in it so it smells of Parker's perfume. Basically, says Semoff, this is *Lovely* thickened with hydroxypropyl cellulose.

Parker: "I like it, but it's been done in Chanel Elixir, and it has a coldness to it."

Semoff shows her a different concept, a rollerball that administers a liquid that could, he notes, also be sprayable—lower viscosity. She tries it, puts it on me, rubs it on her arm, says, "Mm . . ." a bit darkly. Too greasy.

Semoff hands her a silicone substantivity polymer in alcohol. "Here's a silicone formula," he says. "It's called Silicone Polymer 1401." Parker rubs it, smells it. Her eyes fix vaguely on some point where the glass walls meet the ceiling; she's focusing on the feel of her skin. She says, "Mm," but very differently; she's got a faint smile at the feel of this.

"Mineral oil," chants Semoff, "petrolatum, esters, humectants."

Parker snaps into verbal analytical mode. "There are several things I like about the Silicone 1401," she begins. "One is that it's new. Second, that a little goes a long way. Third, is that it really lasts. See? Here it is." She touches the spot on her arm.

Semoff: "You can feel it."

Parker: "I can. Right here."

Semoff: "It's been used very successfully in hair care, but no one has really embraced this ingredient full bore for skin care." Parker puts some on her fingers and rubs it into her long wavy hair.

Semoff: "You can push it in terms of the skin feel, the texture, the imagery you're trying to create. Most people are pushing it in terms of skin protection—that's the heavier silicone polymers, dimethecone and phenyl trimethecone."

Walsh: "Whereas this is really about the scent."

Semoff: "It's really a way of delivering scent that no one's ever done." He looks around, blinks owlishly behind his glasses. "And that's it." Presentation finished. Everyone applauds.

Parker: "Wow. And you have no idea how good you are, making me focus on chemistry. My high school chem career"—she gives a bleak look—"*not* good."

Semoff: "I started in the perfume industry washing beakers in the lab."

Parker: "And when you say 'beaker' to me, it brings back a flood of bad memories."

Walsh says intently to Parker, "We love the idea of silicones— we simply want to create a package that will not read 'ancillary' and not 'traditional perfume' either because the consumer mustn't be surprised."

They talk about glass rods versus a dropper versus a stick. "A glass rod might be really pretty," says Parker, considering it.

Walsh: "Whatever ideas we come up with we'll pass on to Chad [Lavigne, the packaging designer]."

Parker: "How about etching?" She means on the bottle. "Is that terribly expensive?"

Timiraos: "Actually that's not so expensive, and"—he glances at Walsh; he likes the idea—"I think it's a good way to get across to the client the feeling of skin."

Semoff: "What you could do is create a product where you increase the viscosity so that it falls right in the middle, between perfume in alcohol and a gel."

Walsh says to him, "Anything that you can think of that relates to skin is where we want to go; that's what makes the story unique." She pauses. "We can put things in it." She's not really sure she should bring this up.

Parker opens her eyes wide. She says in a thick European accent, "Flecks of Svvvvaroffski crrrystals!" spreads her fingers, and wipes the hand across the space in front of her: ta da. She says to Walsh, "Like back in the day, 1970s—how sparkly we all were. Ohmygod, I don't even like to remember!"

Walsh: "The white eyeshadow that was so reflective!"

They're laughing hysterically. Parker says, "I'm amazed we didn't accidentally blind drivers." She pretends to slouch along a street, turns like a supermodel, and blinds an imaginary car.

Meeting's over. Everyone starts gathering papers, getting up. Walsh suddenly realizes: They haven't had time to get to Perfumery 101. She shrugs. Next meeting.

CHAPTER 8

W HEN ELLENA'S DAUGHTER, Céline, was seventeen, one day in the car as he was driving her somewhere in Grasse, she said, "*Voilà, Papa,* I want to do this métier." She wanted to be a perfumer.

Ellena was quiet for a moment and then said, "Think about it." She would speak about it from time to time. When she brought it up again at nineteen, he knew she would not be convinced otherwise.

Céline's decision left her father decidedly ambivalent, principally because of the psychological difficulties of the life. For every brief he won, there were dozens lost, and each of those was a creative work that died. It was often agonizing. Céline herself would recount later the impact it had on her as a girl seeing her father and grandfather returning home battered after having lost a project. Jean-Claude would struggle months with juices to sit in meetings and be told, "What the hell did you think you were doing?" And "What is this crap?" And, concisely, as a dismissal from the room under many pairs of eyes, "This thing you made—" (a curt, derisive motion) "it'll never sell." They gave the briefs to other perfumers. His daughter watched the fallout. All the brilliant, labored-over ideas he had, the precious works of art that were killed, all the scents he'd borne, then lost, gathering dust unpublished in his notebooks. For a period of time, she knew, he took pills to help him sleep.

"In this profession," says Céline, "people don't talk about it. No one talks about it. But I've seen it. In your head you think: 'What will I say if they don't like it? What will I reply? Because I have to win this brief. I have to. What must I do to win?' You submit your scent, this scent you've worked so hard on, slaved over. People put the *touche* under their nose. And they can be terrible. Terrible. They throw your juice on the ground; they sneer; they take your perfume and throw it across the room."

A few days later Ellena said to her, "You want to be a perfumer. OK, you will be a *laborantine*, a lab tech, at Givaudan over the summer." (He was *chef parfumeur* there.) She worked for him and for other perfumers, weighing their formulae. "I cried," she says. She bursts out laughing at the memory. "The very first project, which a woman perfumer gave me, was to weigh a very old formula. There were almost three hundred materials. An experienced *laborantine* would have spent an entire morning on it. It took me three days. I made mistakes everywhere, too many milliliters of something, or I put in the wrong material. I finally went to see the perfumer, and I told her I'd ruined it, and she scolded me very severely, and I went into the bathroom and cried. I found it so unjust, so difficult."

She adds, "I never told Jean-Claude. He must know. But we never talked about it. I just gritted my teeth and worked. I was too proud." She stops, smiles, says, "I am too proud."

She saw perfumers forgotten, pushed aside all the time. "There are terrible periods where you have nothing to give, no inspiration; you sell nothing; everything you do is rejected. I've seen many perfumers debut and then so many disappear. The profession eats you, grinds you down."

If he was somewhat wary of her having chosen this life, his advent at Hermès gave him an opportunity to help her through it. Ellena had, with a designer friend, created a tiny niche perfume house called The Different Company, for which he had created all the juices—four of them—and when he accepted Gautier's offer, he and the designer had had a brief discussion, then announced that the new in-house perfumer would be his daughter, Céline Ellena. Along with a perfume house, Céline inherited three things from her father: her looks—though Ellena and his wife, Susannah, have donated an exact fifty-fifty split there; her face broadcasts with amazing fidelity her father's slightly closed, slightly dark Mediterranean handsomeness and her mother's open, clear Celtic features—her talent, and a capacity rare among perfumers for talking about perfume. Like his, her words are deliberately but not laboriously chosen, the sentences fluid and given a subtle polish.

When Céline was announced for the position, there was the usual less-than-charitable speculation among perfumers about the ratio of her talent to her name. Then in 2004 she produced for the collection *Jasmin de Nuit*, Night Jasmine, which is a deeply solid example of the best of its category (floral), and, two years later, *Sel de Vetiver*, Vetiver Salt, which is, arguably, the very best of its (unclassifiable). It is an extraordinary perfume.

I called Céline one afternoon when she was at home in Paris to talk about the creation of *Sel*, and the following is not far from a word-for-word transcript of what she said. "*Jasmin de Nuit* was a wink to my father, who loves jasmine. I used the scent of the jasmine we have at our house in Grasse. *Sel de Vetiver*, on the other hand, was born from a dinner I had in Paris and a carafe of water I drank there. A Frenchwoman of Algerian origin named Ferouz Alhali, she loves Africa, has traveled there quite a bit, and I was at dinner at her house, and she served cool Paris tap water in a carafe. But she'd put an herb stalk in it to macerate, a piece of vetiver root. She had seen them use these roots in water like this in Africa, and she'd brought a few back. She actually didn't know what it was. She said, 'I brought it back from Africa, and everyone drinks it and enjoys it.' I drank and ate, I went home, and days later I found the taste of this water anchored in my memory. This was four or five years ago. I had finished *Jasmin de Nuit,* and we were thinking about new ideas. I decided I wanted to do a perfume that spoke about salt. There are many sweet perfumes. None salt. I didn't want marine notes. I had in mind simply crystals of salt, *sel de Guérrande,* from a region in Bretagne, one of the best salts in the world. I understood that this water I'd drunk at Ferouz's apartment, which I now was making at my own house with the Paris tap water—the salt was the thing that created the link between the tap water and the vetiver.

"Paris water comes from natural springs beneath the city, and the taste of the water changes depending on which arrondissement you live in." (Paris is divided into twenty sections called *arrondissements.*) "The water in Montmartre, the ninth arrondissement, is particularly good. In the first, tenth, fourteenth, and fifth, slightly less, although still good. It's not particularly salty, although when it's very hot I find the water becomes saltier, and in winter I find it more mineral, which is a better taste. The trick was going to be finding the scent of salt, which is very abstract because salt is a taste, not a smell, although when volatilized we can smell it. I asked Jean-Claude, 'What is the odor of salt for you?' I asked other people. And every time, depending on the people, I got different answers. People confused the odor of the sea and the taste of salt. I didn't want sea. Or ozone. For me, salt is the taste of skin. When you've perspired, or when you swim in the sea and the water evaporates and you have the salt still on your skin, that is the smell. *La saumure,* when you put olives or anchovies in salt to conserve them. And cabbage, they used to conserve cabbage on ships with salt. I wanted to find some natural materials that were a little bit forgotten, used by perfumers in the generation of my grandfather, Pierre Ellena, who was also a perfumer, *grassois*. He was born in Italy of a family who were probably Greek; they immigrated to France in the 1920s as manual labor in the flower harvests. He often used an essential oil of *la livèche*. He started as a worker in a Grasse distilling factory, and he climbed the ranks bit by bit to become a perfumer.

"I'm a perfumer at Charabot, a perfumery materials maker. It's still a family-owned company after two hundred years, and Charabot specializes in natural materials. I have a studio with a small team in Paris, but the main factory and the headquarters are in Grasse. I went there and looked on the shelves of the factory, and there were all sorts of materials that were literally gathering dust. I smelled *mousse de chêne*, oak moss, which has an algaelike smell, but it wasn't right, and besides it's well known. There was a material that smells of nuts, and I wrote it down in my little notebook that I always keep with me, but it wasn't right either. I saw the *livèche*, and I didn't know what it was. I smelled it, and I thought, That's it . . . *Livèche* smells very spicy, of curry, and pickles. And above all it smells of *la saumure:* salt, water, and food. It smells of skin that has taken salt. To make my salt scent, I took the *livèche* and I added synthetic salicylates: benzyl-, methyl-, amyl-, hexyl-, *cis*–3–hexenyl salicylates. I also used prenyl salicylate, ethyl salicylate, and phenethyl salicylate, which smell very fat and very mineral. It's those that give the perfume's rhythm. If you use only naturals, a perfume is very heavy and inert. Synthetics give the rhythm to a perfume, *le creux et le plein,* the empty and the full coherently together, the rounded whole. The synthetic does"—she sings a single tone and holds it for a moment, looking straight ahead. "Naturals bring light and infidelity. Naturals have several personalities, they tell you different stories, sharp, low, fruit, shade, bright, and they change with time and their mood. The natural does"—in a clear, sure voice she sings several notes, "Dah . . . dah . . . ," some higher, some lower. "When you use naturals, you put in a little bit of synthetic, because with only naturals, the perfume is *très bavard*, talkative, full of little stories it tells. Like when you open your window and you have Second Avenue, all that noise." (She'd asked where I lived, and I'd told her that my apartment was on Second Avenue and that I was closing my window because the June traffic was too loud.) "When you put in synthetics, it becomes an ordered series of sounds; they come together in a key, with rational tones.

"And then a few other materials to trouble the perfume, because a perfume must not be perfect. It must have small imperfections that give it tension. Tiny lights that glint. Which is why I put in a touch of ylang-ylang, which puts in fat, geranium for its lovely metallic and its tiny taste of water. I do very short formulae. Then a natural orris resinoid, what is left over from the highest-quality iris root butter, which has an excellent odor of salted caramel. And you're going to laugh, but it smells a bit of popcorn, but no vanilla, no sugar. I used it for the gentleness. It's a patina, like an old well-polished wood. It's heat, actually."

A child cried in the background over the phone, and I heard Céline gently shut a door. "There are synthetics that give the scent of big, gray clouds and of rain." She thought about this for a moment.

Recently, her son Enguerrand asked his mother for the smell of strawberry with mint. "Like many French children, he loves the scent of strawberry. It's the reference *gustatif* of childhood in France." Then he asked her for popcorn, and when she gave it to him he found she'd done it poorly. Then he asked for the scent of spring. "He doesn't know," she says, "that I used to ask these scent stories of my father."

She turned back to the perfume. "I used two essences of vetiver, both from Haiti. Both are very expensive. One is traditional; the other is *une fraction*, which means it is a piece of an essential oil that one has isolated from the original whole with a machine that slices apart the molecules. It's a molecular distillation. I mixed the two until I obtained exactly the odor of vetiver I wanted. Usually vetiver is quite harsh, it smells of *toile de jute*—the jute sacks you put potatoes in—plus fresh, raw peanuts."

I told her that I don't like vetiver. I've tried and tried. But I find *Sel de Vetiver* essentially perfection. "The vetiver is a pretext," she said simply. "The vetiver here simply helps me to speak of salt. I wanted to talk about salt. *Ce n'est pas la peine de parler d'une histoire que beaucoup avant toi avec du talent ont déjà racontée.*" It's not worthwhile to tell a story that many with talent before you have already told.

From her grandfather, Céline had learned to smell. He would hold her hand and take her into the garden, and they would lean into the flowers. From Jean-Claude, she learned the notion of time. "I am by temperament very impatient, and he taught me about putting time into my perfume. And something else. One day I gave him one of my formulae. He studied it for a moment. He handed it back to me and said, *"Ne pas faire plaisir à son ego et faire plaisir à la formule."* Don't please your ego; please the formula. He handed the formula back to her. He hadn't mentioned a single material.

She wrote it down, and she still has the piece of paper.

⁓◎

On June 14, Ellena went back to Paris for meetings with Gautier and Dubrule. He'd brought up the four new *essais*, AJ1, AJ2, AJ3, AJ4, the next iterations of the perfume, which he'd done in Grasse. At the moment he was hanging out near Dubrule's office in Pantin, smelling them from time to time "to live them" and plotting his strategy.

He hadn't actually decided if he was going to show them to Véronique Gautier, whose schedule no one could quite figure out, and to Hélène Dubrule, who was (possibly) leaving for New York the next morning. "You have to know how to wait," he said, thinking it over, and added, "That's my psychology of the bazaar."

The two executives had been sending him signals. Ellena found that the signals were helpful—and he found that they weren't. This was par for the course in the dynamics of the creative director-marketer-perfumer relationship. Gautier had recently called him in Grasse—she'd been smelling AG2, thinking about it in Paris—and told him to "raise the volume a bit." He had a few methods for doing this that he was mulling over as he sat in the corridor. He could go a bit more into woods by using a powerful synthetic called Karanal. The problem in his view was that wearing Karanal was like slicing your arm open with a knife, olfactorily speaking. Or he could go to the fruits and get a neon fruit, but he *really* didn't want to do that. "They stink. Or they can." Oily, greasy, sticky. So, what then.

There were flowers he could try: lay down a floral track on top of the green/fresh, mix in a bass line in narcissus or hyacinth, but . . . well, they'd have to be neutral flowers adapted to the perfume rather than forcing it to adapt to them. Jasmine? No, he'd be, as he put it, "telling a different story." So "floral abstract, maybe?" he said, looking morose. He brightened for a moment. Maybe an angle starting with lilial, a synthetic watery floral. Maybe a floral-wood accord. He seemed hesitant and pensive. "But I'm not worried. For the moment I'm floating, I'm smelling. Maybe the answer is already in the perfume. Often you just need to turn up something already in there."

But Ellena was dealing with something else. He was well aware that beneath these four vials lay a vast, difficult question they would have to answer, relatively soon: What was this thing called an Hermès perfume? If you are Stella McCartney, you do not worry about history and tradition. You launch *Stella*, which is a solid-enough piece of fashion perfumery, a smooth, delectable darkish rose that is slightly powdery, and you are blissfully unburdened by a century or so of a house's history, that history's aesthetic, and how to put that aesthetic in that fragrance. If you are the Olsen twins, you meet your agent at the licensee's corporate offices, sign the contract, and walk out of the room, and your licensee eventually launches two scents that smell like car exhaust on Tenth Avenue, scents with no *persistance,* no *sillage,* no beauty, and no reference to anything except its creative team's attempting to ride some vapid pop cultural pulse. They do the marketing for you, and you cash the checks.

If you are Hermès, these solutions are anathema. It is, said Ellena in his corridor, a privilege and a burden to be Hermès. Regarding the burden, Gautier knew all the criticisms, and Ellena knew what people said: That Hermès's perfume collection had no unified signature, that nothing linked *these* scents to *this* house. That Hermès had never really put the same effort into its fragrances as it did into its other products. That Hermès had flirted with fragrances but had never fully committed.

To this, of course, Gautier had a reply, and the reply was Ellena. He was the commitment. But he was conscious that this almost two-century history was now sitting on his shoulders, and it was he, not Gautier nor anyone else, who ultimately had to find the solution to the question of what made a perfume an Hermès perfume.

Arguably the strongest signature of any collection currently on the market is owned by Armani. L'Oréal, which owns Armani's fragrance license, has done a positively German job of precision engineering every single launch under the brand—from 1995's massive bestselling feminine *Acqua di Giò* (created by the two legendary perfumers Edouard Fléchier and Françoise Caron) to the huge 2002 hit *Armani Mania Pour Homme* (by the young wizard Francis Kurkdjian) to the slightly robotic *Black Code.* Each time at bat, L'Oréal has ensured that the perfumer weaves in the same filaments. Brand unity. You smell the links subtly but distinctively, not as materials but as style, the juice olfactorily finished in that instantly recognizable matte, sleek, silver-gray Armani polish. An aluminum carapace, one part light to two parts dark, and the perfumers manage to convey it in smell. Even the brightest Armani scents you view as if through sunglasses.

I don't like the Armani aesthetic particularly, and I never have. (For some reason I believe in Armani-ness in the clothes, where the sleekness translates into elegance, but not in the perfume, where it strikes me as freezing-cold modernism.) But I admire it intellectually, and I respect how Armani's creative director, Karine Lebret, has hired and ridden herd on the perfumers, extracting from them the technical mastery necessary to embed into each juice a strand of Armani's DNA, this scent signature, as invisible and powerful as a silicon locater chip. From a marketing perspective, it's genius. Other brands have, as brands, less coherence. Yves Saint Laurent's *Paris* is one of the most gorgeous and elegant perfumes it is possible to buy, arguably the paradigmatic rose, glorious and luxurious and of excellent quality, and Yves Saint Laurent's *Baby Doll* is also brilliant, arguably the paradigmatic scent of helium-induced laughter and ketamine pinkness, territory the Fiorucci scents pioneered in commando boots and tutus, but what does *Paris* have to do with *Baby Doll*? And—strictly speaking—what does *Baby Doll* have to do with Yves Saint Laurent? Individual scent by scent, I would in almost all cases prefer an Yves Saint Laurent perfume over an Armani, but when it comes to the unified aesthetic, the brand signature, Armani is spotless.

When the signatures work, it is, generally, because they have been carefully engineered. The Escadas have a well-developed, palpable signature because their licensee, Puig, does an excellent job of having the perfumer place the same olfactory thumbprint onto each Escada scent, a sort of neon-and-disco-ball party-crazed Barcelona-Eurotrash-girls-gone-wild aesthetic. It works beautifully in *Ibiza Hippie*, a scent that incarnates the smell of the taut tanned neck of the girl at the beach that everyone wants to dance with— sea air, sweet beach foods, cotton candy, plus the traces of suntan lotion in her sweat. And less beautifully in the hideous *Escada Magnetism for Men*, a mixture of Papaya Kool-Aid and *Brut* by Fabergé. Estée Lauder (which owns the license) has kept a similar firm hand on the controls at Tommy Hilfiger, whose signature is a quintessentially American openness that feels both comfortingly familiar and (reasonably) avant-garde.

Coty—and Kenneth Cole, per his licensing contract with the company—makes millions off the Cole signature, mostly in the Midwest where the Cole scents are wildly popular with middle- and lower-class straight guys who think they are buying the olfactory equivalent of a pair of Kenneth Cole shoes but wind up smelling like fluorocarbons. I smell Kenneth Cole, I think of the hole in the ozone layer. A Coty executive once said to me, "You can't argue with money." Yes and no: Kenneth Cole scents smell like air-conditioning vents.

The deepest pit of hell, however, is reserved for the suits slaving over the Hugo Boss brand in the dark corners of Procter & Gamble, which owns the house's perfume license and creates its scents. In the St. Petersburg airport, where I once spent an hour waiting for a flight to Amsterdam, they had every Hugo Boss. P&G has created for Boss a fluorescent-lit polyurethane signature that is as clear as it is repellent. The fact that P&G sells more Hugo Boss–scented products to men on the planet than almost any other brand contradicts the economic theory of the Invisible Hand (which is supposed to create a certain logic) and confirms Veblen at his most bitterly cynical. I know someone who believes that Hugo Boss scents constitute proof that God does not exist. I would disagree—atheism is a rational response to the chaos of life, while the rational response to any perfume by Hugo Boss is to run, screaming, into someone's parking lot—but the scents certainly do make you think of Agent Orange. Here are a few of them.

> *Soul.* This smells like the very good, very interesting *John Varvatos* except that everything good and interesting and creative has been leached with chlorine from its desiccated corpse. Given that it has none, calling this smell *Soul* is like calling Sean John's hackneyed, derivative masculine cliché *Innovative:* It is an Orwellian use of the English language. This is the chemical

reek of deep-space travelers frozen in goo in suspended animation.

Hugo Boss. Huh?

Edition. Grout cleanser.

Intense. Nice, sort of, if you don't really want to smell of anything at all, until it fades into complete obscurity in about four minutes and disappears into some dimly lit service corridor below an anonymous sewage treatment plant.

Natural Spray. This scent makes us ask what alien species does the marketing for Boss. The brand has a phenomenally irritating tradition of not naming any of its scents coherently, or of intentionally obscuring the name because apparently on this species' home planet "obscure" is what passes for "cool." (Hey, what's this? Who knows.) (Also, most of the juices look like frozen green goo.) As far as I can figure out after a close perusal of the packaging, *Natural Spray* is this thing's name. It could be sold in the local drugstore next to the disposable diapers. Some experts think aluminum induces Alzheimer's. This stuff smells like aluminum—experience it by holding under your nose a cheap metal spoon with traces of dishwasher detergent chemicals. After you smell it, you may desperately order an eight-course all-aluminum tasting menu with an aluminum chaser in an attempt to induce a case of Alzheimer's sufficient to forget the smell. At the same time, to describe this perfume is to render it more important than it should be. It is nothing.

Elements. A cologne most appropriately worn by electrical appliances. It should be called *Eau de Refrigerator Condenser Coil.*

Number One. Fascinating. If a cat had morning breath, then ate kibble, then licked its anus, then licked your hand, and if you then smelled your hand, it would smell like this. (Number One what, incidentally? The number one nuclear meltdown in a paint plant? Or maybe the one that's not number two. If the name of this perfume is not *Vile and Toasty*, it is only due to the lack of the appropriate application of genius in the Hugo Boss marketing department.)

Woman. This is a scent for a woman who has no taste and absolutely no interest in having any. This is for a woman who loves a man who wears *Vile and Toasty*. She lives in an apartment without light or furnishings or running water or, perhaps, oxygen, and sleeps inside an aluminum container in a frozen green goo.

In a sense, you have to admire P&G's Teutonic monochromaticist steamroller. It's able to stand up to anything. That thickness of armor plating demands respect. And perfume-as-amphibious-assault-vehicle is an interesting aesthetic (*Natural Spray* is more of a flame-thrower), but it's about as pleasant and as subtle as Kevlar. The fact that Boss and P&G have managed to wring such utterly wretched products from such considerable talents as Olivier Polge, Alberto Morillas, and Dave Apel merely speaks to the iron-clad strength of the Boss signature. The creativity of these estimable perfumers is simply ground out of them under the steel fist of some executive with the soul of a creature from a Bosch painting, who is himself just driven by the market even as he drives the market toward the core of a nuclear reactor. The brand's signature destroys everything before it like the clouds of nerve gas that drifted silently and murderously across the trenches of World War I, poisoning every living thing they touched.

By contrast, the central aspect of Jo Malone's scent aesthetic is, in my opinion, the quality of light. When I mentioned this to her over breakfast at a Manhattan restaurant, she said, "Hm!" Then she paused, considering my remark. "I've never thought about it as such."

Really? I said, surprised.

"Well, no . . ." And then, half laughing, half in irritation, "Well for godsake, Chandler, don't look at me as if I was an idiot!"

Sorry.

"I mean," she said a bit uncertainly, "it's very *interesting* . . ."

And I was thinking, But isn't it totally clear? Malone's work is criticized as "aromatherapy" and praised as "contemporary." Her genius in either case is the fingerprint she leaves on each scent, a marvelous quality that is not weightlessness—it's something much more startling: Weight that floats, hovers in the air. Solidity shot through with light. The classic Hermèses are thick palpable fabrics, perfume as rich, opaque silk weighty on your body; Malone's are like the revolutionary new-technology translucent concrete that architects have just begun using, a recipe of glass gravel mixed with optic fibers. When poured, this concrete forms slabs of luminescence, and the outlines of people inside the walls of buildings using it are visible to those outside at night against the glow of the light inside.

Think about Malone's *Amber & Lavender Cologne*, the *Grapefruit Cologne*, the *French Lime Blossom.* I mentioned to Malone the "glass roof sensation" I feel when I smell her work, and she frowned.

"In my perfumes?"

Yes, yes, all the light, I said, and hesitated; we seemed, suddenly, to have missed each other. I asked her, When you say light, do you mean light, the antonym of heavy, or light as photon radiation?

"Oh," she said immediately, "the antonym of heavy."

Ah. I'd been thinking of luminescence, she, of heft. She cocked her head for an instant, pulled a small bottle from underneath the starched white linen tablecloth. "I'm not going to tell you its name," she said. "It's our next launch. But you tell me whether you find your glass roof." She sprayed it on my arm, sat back.

You bring the arm to your nose, and you think, Wait . . .

What is light to you? This perfume is the scent of the darkness that inhabits the corners of the paintings by the Dutch Masters.

You know Rubens's *Self-Portrait?* The rich, rich, purple, luscious dark that surrounds that illuminated head watching you, that bright white collar floating in the warm blackness. Rubens's dark is not the cold heaviness of the void. It is the deep warmth of all that is there but, simply, unseen. That is this scent. It's entirely possible: Think of photons, particles with energy and momentum, yet, quixotically, no mass at all. They're all around us. What is amazing about this fragrance is that it is at once utterly different from Malone's bright work and, yet ("Oh! the antonym of heavy"), characteristically massless. Jo Malone, queen of light, has created the weightless scent of a photon.

⟁

Ellena was thinking about all this. He had to resolve the problem of the Hermès signature by July. He was aware of the views out there.

Over a salad in a sixth arrondissement café, a rather severe Parisian woman long in the industry described Hermès perfumes to me as compact. Handsome. Forbidding. *Maîtrisé*, masculine, the opposite of fantasy, never, she stated firmly, a *Fracas* by Piguet, this gloriously over-the-top tuberose, never the mad gourmandise of Mugler's *Angel*. They were not easy, these Hermès fragrances, she said, stabbing a fork deep into the flesh of a Bibb lettuce. Not accessible. Not feminine in any common sense. She took a sip of wine, put the glass down decisively, and made two surgical observations: "They"—she meant Hermès—"like animals more than people." And: "*La fragilité n'est pas leur truc.*" Fragility is not their thing. Hermès perfumes were almost barricaded; look at *24 Faubourg*, she said. "*Verouillé à tous les étages.*" Every floor locked tight.

If perfumers defended the house—"The Hermès signature is quality," said one "elegance, good taste, quality, you smell all the expensive ingredients, you see every dollar on the screen"—mere "quality" as an aesthetic left something to be desired. Bacon's paintings are "quality," and so are Caillebotte's, but the signature problem turned on the fact that the eye would never mistake Caille-botte's gorgeous bourgeois Parisian genius for Bacon's slide shows of blood-spattered hell, nor either of those for David's neoclassical clockwork beauty. If Hermès's and Chanel's scents were both qual-ity, in a perfect world an Hermès perfume would never be taken for a Chanel, and while consumers expect quality, they hunger for style. And style is, substantially, signature.

There were obvious paths Ellena could follow, but they were not necessarily workable. Hermès was leather. The house began as leather workers making saddles. So perhaps a subtle leather note in every perfume that said, "I am Hermès," that lent coherence. But if the leather scent in *Bel Ami* was indeed quintessentially Hermès, did they just renounce the citrus in *Orange Verte*? And how did you make leather friendly to the twenty-first century without falling into S&M? Ellena had mentioned the leather idea once or twice in pass-ing, but for the moment he was simply eliding the signature ques-tion. He, Gautier, and Dubrule had decided that, for the time being, the Hermès signature was to be luxury. A fluid concept, but one would be supposed to know it when one saw it.

Dumas had hired Ellena to implement the signature. Ellena was conscious (the Mona Lisa smile again) of coming aboard the house just behind bad-boy/iconoclast/sexual provocateur Jean Paul Gaultier, whom Dumas had in the fall of 2004 picked as de-signer of Hermès's women's collections. For staid, conservative, very Right Bank Hermès, the Gaultier move was much discussed. "I admired Hermès till they hired Jean Paul Gaultier," sniffed one doyenne of Paris's luxury industry. "I was astonished that they took him." Others saw a savvy and necessary makeover, a move toward *le commercial*, aka products that actually moved. A young New York fashion editor gushed to me that "Hermès is doing, like, a total Burberry!" (the fading British house had revived spectacularly), but it was a risk, and many in the fashion world were not certain that Hermès should in fact be doing a total Burberry. Here was a house whose products spoke, metaphorically, to Proust's Madame de Guermantes. Gaultier was more likely to be found chatting with Paris Hilton, who, incidentally, had just come out with her own fragrance—"Your chance to share the magic that is Paris Hilton," said the press release—the juice built for her by the perfumers James Krivda and Steve DeMercado. That perfume was a perfectly well made, super-commercial neo-luxe scent, but it did not bring to mind words like *taste, quality,* and artisanal *craftsmanship,* and Macy's sold out of it twice before it was even officially launched. To Ellena and Véronique Gautier, this served as a bracing splash of the reality of the harsh new perfume environment in which they were operating. Gautier wanted Ellena to create art, but she wanted the stuff to sell too.

Ellena is actually a bit of a diva about *le commercial.* "I'm certain that AJ2 will be the most commercial," he said, then added very quickly, "but I *never* take a position on the market. I give my opinion on the aesthetic. When they start asking me for a commercial opinion, I start braking with both feet." Ellena said he only smelled new launches once because he didn't want "to be influenced by the market." He might look around for things he could pirate, "but, you know" (this somewhat primly), "there's not a hell of a lot on the market worth pirating."

So here he was, sitting in a corridor in Hermès's offices, smelling his *essais.* "I'm pretty happy with the *sillage* in the latest versions," he remarked. "The problem is the *persistance.*" He stopped. "But I'll find the answer." Stopped. "I *hope* I'll find the answer." Stopped. "I *must* find the answer." He smelled. His face went agonized.

He said thoughtfully, "You know, perfumes have a certain threshold in terms of the number of bottles sold, above which no perfume will ever go. Because when it reaches a certain concentration, people start turning against it. It's not like Coca-Cola. Which billions of people can drink."

Dubrule and Gautier were, for the moment, still nowhere to be seen. He remained undecided. He looked at the vials. "I still don't know if I'm going to present them or not." He considered.

He felt blocked, at moments. Sometimes, he remarked, he would sit at his desk and . . . nothing. And nothing. He screwed up his face morosely, then broke into a huge smile. "They asked Picasso, 'What do you do when you're inspired?' Picasso said, 'I work.' They asked, 'What do you do when you find yourself blocked creatively?' Picasso said, 'I work.'"

He smelled the vials again. "Black skin is the best for perfumes. It retains fragrance much better."

He smelled them one more time. "Now I like AJ3. The mango. It's fuller. It fills out better." He peered at a vial, made a pensive moue. "The problem with AG2 is that it's too citrus." He looked subtly stricken.

※

The next day, on June 15 at 10:00 A.M. in Paris, less than four weeks from the July 10 deadline for finalizing the perfume, Ellena arrived for a meeting at Hermès's luxurious headquarters at 24, rue du Faubourg Saint-Honoré. Hermès's flagship store sits on the street level, under the headquarters, and down from Chanel, Dior, Missoni, and the Hôtel Costes. It is also a block from the Hôtel Crillon on the place de la Concorde and the relentlessly, aggressively hip Buddha Bar, a restaurant that finds itself, perhaps to its surprise, increasingly barricaded behind the elaborate security measures implemented by the American Embassy directly on the other side of the rue Boissy d'Anglas.

There are usually uniformed military personnel with machine guns in the street here. Due also to the presence of the British Embassy four hundred meters to the right of the store and le Matignon, the French president's home, a few hundred meters farther from that, this block of the rue du Faubourg may be the most highly militarized superluxury spot on the planet.

The beautiful summer mornings were stretching themselves into a surprisingly major series, and Ellena and Anne-Lise Clément, the project head for perfumes, walked up the narrow sidewalk past Lanvin under blue skies and warm yellow sun, police discreetly everywhere.

It was Ellena's first meeting in this august place. They arrived at the elegant 23, rue Boissy d'Anglas business entrance, a spacious arch of stone built for eighteenth-century horse-drawn calèches. At Hermès, even the business entrance was relentlessly exquisite. The bright sun made the space inside the stone arch pleasantly dark and cool. The Hermès guard handed them a visitor's card to fill out. Ellena peered at it. "We need to complete this?"

It wasn't his words but his manner that made the guard suddenly frown, already shifting gears. "*Ah!*" asked the guard, a shade more deferentially, "*mais vous êtes avec la maison?*" But you are with the house? Ellena, just perceptibly, straightened up. He nodded. This was the first connection, and from now on the guard would remember his face. There was, instantly, a complete attitude change, the standard officious French attitude now an attitude of officiousness that included Ellena. "*Mais vous montez tout de suite!*" said the guard. But you go up immediately!

The sterile fluorescent-lit service corridors inside are identical to those backstage at Saks and Bloomingdale's, the harsh lighting, the clean, elderly, comforting formica department store floors. A young woman arrived to collect Ellena and Clément, led them inward. They passed for an instant by the store's back door, glass covered with a black lacquered iron grille (so different from this utilitarian taupe), a view into the lush, rich interior of Hermès filled with the vibrant, bright colors of the towels and plates, the shine of the silver objects, rich leathers. They walked, bumped up a few steps, down a few steps, and were lost in twisting, turning wormholes that disappeared here and there.

Ellena grinned. "*This* is why things are slow at Hermès. You have to find the right office. You could get lost forever here, you starve inside these tunnels, and they find your body in the spring." It was a rabbit warren, stairways that communicated in narrow halls. Their guide expertly herded them into an elevator, which sighed upward and stopped, and suddenly, blinking, they emerged into the real backworld of Hermès. Under Paris's rooftops are breathtakingly lovely ateliers. Old French windows, large panes of glass surrounded by wrought iron, let the blue and yellow light into cool corridors. There was a feeling that everything was silk and beautiful and pure. As they passed one of the ateliers, a lovely young blond woman in an exquisite pale-pink top, a secretary perhaps, was delighting in a small leather-and-cloth bag an Hermès artisan had just turned out. He stood behind his bench, a large, thick, meaty Frenchman torn between gruff restraint and open pride, and basked in her delight as she opened the small gold-colored clasp, felt the inside, closed it again. She swung it around and laughed. Her laughter was like music. He nodded gravely at the bag he'd made.

Ellena and Clément were here for a *réunion artistique*, a creative meeting to discuss the small illustrated book about their trip to Aswan that would come inside the special *prestige* box set of *Un Jardin sur le Nil*. Ellena would write the text, photos by Quentin Bertoux, and Dubrule had decided the box set would include sticks of incense perfumed with Ellena's *Nil* scent, so they'd be discussing how to manufacture those as well. Ellena and Clément walked into the meeting room and greeted Fred Rawyler, the tall, thin, serious, fiftysomething Swiss German who oversaw the house's design. Rawyler, elegant and intense, wore glasses and a late nineteenth-century haircut, longish and Beethovenesque. For twenty years he had worked with Hermès. Everyone did double-kiss *bises* all around. Bertoux, the photographer, walked in, and they complimented him on the current Hermès magazine cover photo. It was an exquisite little girl in the paradigmatic sailor's dress of childhood, arms extended behind, chin up, and leaning blissfully into a jeté. She was Bertoux's daughter.

Stéphane Wargnier, head of communications worldwide, arrived last, blew into the room with a big curly mane of hair and a big curly personality. *Bises* all around again. Everyone sat down at a round table. The offices were pale, pale celadon.

Rawyler lay out the galleys of the book, photos and the text, which Ellena had drafted and they'd then translated—English, Italian, Spanish, Arabic, the alphabets of the various languages like different spices. They talked about the layout, contemplating every millimeter, every detail.

Wargnier: "Remember that Jean-Claude's text is independent from the photos."

Ellena, smoothly: "Exactly. It's its own story."

Wargnier was wearing his usual unusual clothes, an amalgamation of styles and bright fabrics and layers.

Bertoux: "Jean-Claude sent me some *flacons* (perfume bottles). We could use images of them to initiate each section." Icons, basically.

Wargnier (to Ellena): "You sent him the perfume and not us?" Wargnier was in fact joking; he assumed Ellena wouldn't have.

Ellena looked at him, realizing: "You haven't smelled the scent?"

There was an instant intake of breath around the room. Ellena's hand moved toward Clément's, hesitated, and Wargnier realized and said, *"Non!"*—almost leaping over the table. He made an impatient gesture: gimme!

Ellena to Clément: "You have it?"

She got out the current draft, AG2.

The room watched in silence as Wargnier sprayed it on. He closed his eyes, waved the arm leisurely under his nose. Someone honked outside on rue du Faubourg, a car passing the British Embassy. He opened his eyes and smiled intently.

Wargnier: "It's not the opposite of *Jardin en Méditerranée.*"

Ellena: "It's the next."

Wargnier: "It's truly the second chapter."

Ellena (smiling): "That's what I wanted."

Clément: "Very different, but you recognize the *Jardins* collection."

Ellena: "And there'll be a third chapter, and a fourth." He said wryly, "The second volume is written by Victor Hugo."

They were passing the bottle around, everyone spraying. Ellena watched them closely, wearing his Mona Lisa smile. Wargnier was frowning at his wrist, breathing in, out, in from his nose: "And it's funny," he said intensely, "because the odor is actually *so* much more *colored* than the garden in Aswan. It's a Rousseau painting. Colors all over."

Ellena: "Very colored."

Wargnier: "Jean-Claude, you watch the volume!"

Ellena: "I'm working on it! . . ."

Wrist still to his nose, Wargnier moved a critical eye over the layouts and photos spread across the table. "Hm. Smelling it, it occurs to me that the packaging design is a bit too demure for the perfume . . ." He jots down a possible title for the book: *Impression du Nil.*

Ellena: "How about *Nez Fertiti*" (*Nez* means "nose").

Rawyler: "Quentin, you could do us a great photo there!"

Bertoux: "Introduce me to Fertiti."

They went over the printout of what Jean-Claude had written. Possible titles for sections: "*Les Mangues Vertes.*" The Green Mangoes. "*Les Saisons.*" The Seasons. What was too short, what contradicted what.

Wargnier: "It's the same format?" He meant the specs on page size and paper.

Clément: "Yes in all dimensions except we can go a bit thicker if we want."

Rawyler showed them a font he was proposing. Wargnier squinted at it: "That's going to float a bit."

They talked about *pavés,* blocks of text, and the matte versus brilliant textures of paper. Wargnier, smelling his wrist while he listened, said forcefully, "Well, if we go by the perfume, we need brilliant." (Ellena found this a nice visual way of describing the fragrance, but it directly contradicted Gautier's latest critique, which he found slightly worrisome.)

Ellena sneezed three times. Wargnier looked at him. "I can turn off the air-conditioning."

Ellena: "No, no!" Getting out some tissues. "The room is nice and cold."

They discussed how they should refer to the book. Wargnier suggested *bréviaire*, then immediately, "But no one knows what a *bréviaire* is anymore!" (A *bréviaire* is a small book monks read containing a thought for each day.)

Bertoux: "It's a story that was in Jean-Claude's head during the trip."

Ellena: "No, the story in my head during the trip was, Where are the crocodiles?"

Wargnier looked at Rawyler. "Is the cover a photo of the Nile?"

Bertoux: "The cover!" He paused, then: "*Une photo du Nil, non?*"

Rawyler, warning them: "You put a photo, you've got a tourist guide."

Wargnier: "A big close-up of Jean-Claude's nose."

Ellena: "Ah, excellent!"

Rawyler, decisively: "I just see green. Period. And what do we put *here?*" He meant Ellena's name on the cover, how to identify him. Wargnier shrugged and said, "*Ah, bah, 'Jean-Claude Ellena, l'Académie Goncourt.'*"

Ellena cracked up. The Académie Goncourt is one of France's most illustrious literary societies.

Wargnier sighed. Leave it for the moment. "It's starting to take form," he said. "Not bad!" He went back to his wrist. "It's a big garden. It's not some little housewife's vegetable plot." Wargnier looked with pride at Ellena, then remembered something. "*La mère Gautier,* she owes us a dinner to celebrate the arrival of Jean-Claude." And then he shrugged. "*Bah!* He's already here now."

꿨

The packaging of a perfume can be as much an obsession as the juice. "The design of the packaging is an absolutely huge invest-ment," a French executive said to me. "There's stock packaging, and if you go that way, you can save some significant costs, it's actually a good business decision, but people are very sniffy about it. The box for *Miracle* cost Lancôme maybe $15,000. BPI's Gaultier boxes are much more expensive."

Naturally there's a certain cynicism, and a certain guilt, when more attention is paid to the box than the juice, and Dubrule and Gautier had no intention of committing that sin. Ellena's elixir was primary. Still, Dubrule would doubtless be producing the most ex-quisite package possible to surround it. It was her responsibility, and she would be judged on the success of the look she produced.

They'd been searching for the visual signature a scent needs to establish its identity and represent the house. Dubrule had noticed, about five years earlier, that Hermès Tableware had created a collec-tion of porcelain called, coincidentally, *Nil Hermès*. The motif was lotus leaves in a delicate red-accented, yellow-tinted green. The artist was a woman named Véronique de Mareuil, so Dubrule had asked her to submit a design proposal for both the *Nil* perfume box in its various sizes and the *coffret* (the box set that would include the book of Ellena's text and Bertoux's photos). They'd demanded cer-tain elements, primarily a white background since that was the way the lotus leaves were presented on the Hermès plates and since, more important, it was strategically important given that the *Jardins* were now a collection: With white, they'd be able to do individual future designs on the same basic *étui*.

De Mareuil had done up a proposal, based on her *gouaches*—opaque watercolors—for the porcelain. After they'd found it a bit large, she had reduced it, but they still weren't totally satisfied. Perhaps a white lotus flower on the box's left edge? Maybe she could move the leaves around a bit. Then she did a full mock-up (by hand, which had surprised them), which they had presented to Dumas, and the elegant old gentleman had loved it.

So at 4:15 P.M. on June 8, Dubrule and Clément had had a packaging meeting on the box for *Un Jardin sur le Nil.* The meeting had taken place in Dubrule's office, where de Mareuil was presenting the scanned and cleaned-up version, though she had again brought hand-done individual leaves. She was dressed entirely in black, just back from her vacation in Greece ("What a tan!" said Dubrule enviously, "me, I'm white as a sheet"), chic cat-eye glasses, and sandals. She spread her painstakingly painted leaves across the tabletop in Dubrule's spacious office, then stood back and looked at them, confused. "I didn't realize this flower was cut." She frowned. "Was this the same size?"

Clément: "No, we grew that one."

Hélène Dubrule was wearing a simple summery cotton top and skirt. She was surrounded by the graphics she'd thumbtacked all over her wall: photos of Hermès scarves and the scarf shirts that Jean Paul Gaultier had recently done for the collection. Sketches of purses. Clément was in linen and sandals, sunglasses holding back her shoulder-length, thick dark-blond hair.

De Mareuil: "Now *this* is great, the side, these three leaves above the *Hermès.*" She pointed at the word, which had been professionally imprinted on the *coffret;* it looked, for all intents and purposes, perfect, ready to ship, but all three women saw problems with it.

The sides and back were disproportionate. "Maybe we should keep the proportions here," suggested Clément, pointing to the back of the *coffret,* but Dubrule shook her head: "Then we have to cut it off. This stem"—she motioned. De Mareuil followed the motion and noticed something that Autajon Coffrets, the manufacturers, had done with her design: "Ah! See, here? The stem ends *inside* the box. I love that."

Dubrule, who hadn't actually been pointing at that detail: "Let me see? Ah!"

De Mareuil tried to lift one of her leaves, tearing it in half. She glowered at it.

Dubrule: "You have to buy repositionable glue!"

De Mareuil: "*Oui, oui, je sais.*" Yeah, yeah, I know. She used a ruler to find the center, distributed leaves, moved them lower, marked almost invisible parameters in sharp pencil. They worked efficiently, rapidly, speaking in quiet, focused voices.

De Mareuil found herself struggling to manipulate tiny bits of paper stem, and they burst out laughing. She put a circular lotus leaf with a triangle cut out of it on the box. Laughing: "*Mon pauvre dessin!*" My poor design.

Clément left the room, came back with a paper mock-up "twilly," a long thin silk scarf that is a signature Hermès product. Hermès would be making twillies as launch presents for reporters and for the saleswomen to wear when they were selling *Nil*.

Dubrule raised an eyebrow at de Mareuil: "Twelve *cadres.*" She meant the number of passes the silk dyers would have to make to produce de Mareuil's design. She cast an almost involuntary guilty glance at Hermès's atelier across the street, where many of its products—silk scarves and ties, bags, leather goods—are made. De Mareuil acknowledged (in the breezy, matter-of-fact French manner that never admits fault, particularly where beauty is involved) that this was her fault. Yes. The cost of art.

Dubrule: "And you're going to bend this lotus leaf up so we see the bottoms."

De Mareuil: "Like these."

Dubrule: "Yes."

De Mareuil: "Yes."

They stood back.

Dubrule: "*C'est bon.*"

De Mareuil: "*Oui.*"

They discussed the problems of getting the same colors in the photogravure, making the master impressions. Clément ran down the technical details. They talked about whether the twilly's background should be white or a bit ecru. Dubrule looked confused: "We were talking about a green background."

De Mareuil gave a look of starched horror like Jeanne Moreau playing a schoolteacher: "*Ah, non! Le vert, non!*"

Dubrule acquiesced. She surveyed the revamped design, checked a mental calendar. "How much time do you need?"

De Mareuil: "When do you want it?"

Clément: "The fifteenth of July we want to have the documents at the printers."

De Mareuil: "I'll give it to you around the twenty-third of June."

Clément (reminding her): "And we need the text for the back of the *coffret.*"

De Mareuil: "Jean-Claude is writing that?"

Clément: "No, no, someone else. Jean-Claude is writing the book."

Dubrule: "It's for the presentation for the *produit.*"

Clément: "It can be late July." She put her lips together and said tightly, "We're very late."

Dubrule (grimly): "They're giving you impossible deadlines."

Clément (shrugging): "We have to do it."

De Mareuil: "When does the perfume launch?"

"February, March," said Clément, and she looked at her watch. Then to Dubrule: "We need all the documents to the printers by July to get everything back from them by October."

Dubrule said nothing, breathed in, thinly thinking about the dates. Breathed out.

They stood back and gazed at the green glued paper lotus leaves and the unfinished designs.

❧

Ellena was in Grasse, sitting in his lab office one afternoon, thinking about the approaching deadline, and wondering where he was going with *Un Jardin sur le Nil*.

"Classical perfumery," he said, "is too perfumey for today's sensibility. It's like reading Stendhal. Very nineteenth century." He looked at one of his latest iterations of *Nil*, the small bottle that he held in his hand. "This is a new way to write perfume. This is a new way to write formula. Short, concise. In *First* I put four different jasmines, three roses, two lilacs. Redundant." He dismissed it, then added, "Big bouquets, very complex, very obvious." Dismissed it again.

He looked around for an analogy. "Classical painting is beautiful," he said, starting again, "but it takes so much time to read all the symbols—which saint is it, what does he represent, which biblical story—that it's more about the process and less about the emotion." He mentioned Giotto's frescoes, loaded with elements. "Matisse," he said by contrast, "you just feel it immediately. I couldn't do *First* today. Today I look at that formula, and I think, Oh my God! I was not controlling the materials. And I was trying to understand the market. In *Nil* I know that I mean to put my paintbrush here. And here." He lightly touched an imaginary brush to his desktop. "Not here. But here. I'm not trying to understand the market. I make the market."

One's taste in perfume develops and evolves sophistication like one's taste in music. A person matured, Ellena noted, from, say, simple lilac to the thick, sophisticated, white-breasts-in-satin-gowns-at-the-Moscow-Opera of *Calèche*. It took some effort and some education. (Frankly, getting to lilac can take a little work too.

Lilac smells of soiled underwear.) We all go through the experience of acquiring a taste for certain foods. You start with milk and white bread and eventually reach beer, Gruyère, the suffocating dirty rich weirdness of truffles, pinot grigio. Everything children refuse to eat they pay serious money for as adults, but they must evolve to those choices and those restaurant bills.

Ellena respected the classic French school, of which *First* is one of the later-era examples. (It had less animal but all the gilt and brocade of that style.) He respected, perhaps to a slightly lesser extent, the classic American school (Lauder's *Youth-Dew* and *Aromatics Elixir* by Clinique and *Charlie*, which simply made a more commercial photocopy of the French and put on an American wrapping). But he had evolved. He was not interested in writing Stendhal. If he looked down on the contemporary American school (clean laundry scents like *cK One*), he loathed the internationalist school, the euphemism for the Asian consumer school (Versace's *Bright Crystal*, a piece of chemical-sweet commercial emptiness that smells like a tablet of saccharin). Ellena was in the process of creating a new school.

Not consciously. Naming it would be difficult, if not impossible. The Flesh School? Not that it smelled like, or of, flesh. Its relationship to flesh was the approach. "A good fragrance becomes part of the wearer," Frédéric Malle once said to me; you didn't, for example, want one of those air-conditioning fragrances like Calvin Klein's *Crave*, machines that Malle witheringly described as "those ozonic white-musk gadgets that just sit on your skin."

And yet in a sense the classic French perfumes were exactly that, gorgeous objects made by artists, olfactory jewels set in olfactory fourteen-karat filigree that sat on the skin, separate from the wearer. These were hard, glittering things worn to decorate oneself, and the ego of the artist and the ego of the house came through at a thousand degrees. *Black Orchid* by Tom Ford is the paradigmatic example of this aesthetic re-created today, a perfume that is, intensely, about the perfume; "I am a work of art, see me." By profound contrast, Ellena's scent for Malle's collection, *L'Eau d'Hiver*, melted into your skin and mixed with your bloodstream, and it seemed that you weren't wearing a scent at all. The difference was that with the classic French school, people smiled and told you, "That's a wonderful perfume," as if you were wearing a beautiful new shirt, but with *L'Eau d'Hiver*, they'd get an intense, unfocused look and say, "You smell great," and then stop talking, as if they'd been caught in an inadvertent intimacy concerning your body. It was as if the scent were coming through your pores.

Sake, created by Christelle Laprade for Fresh, is one of this modern school, incidentally, as is, unexpectedly, the crystalline, transcending *Juicy Couture*, one of the best perfumes on the market, by the expert commercial technician Harry Frémont. These perfumes are ideas of the smell of our bodies in heaven.

Ellena was an experimenter. Just as an intellectual exercise, he had taken Hermès's old masculine, the stinking-animal *Equipage*, and redone it in two different lavender versions. One of them was extraordinary and wonderfully strange, the scent of the Thai dish *larb:* lemongrass, basil, mint, and hot red peppers mixed into cold ground pork. You smelled the organic burn and the animal fat. Ellena was uninterested in whacky abstract stuff. He liked perfumes that were *bien balisés*. Well marked. (*Les balises* are the lights on either side of the runway that guide the plane in.)

He was crafting a style, but not everyone understood. "Some people," he said, "smell my latest and say, 'Well, it's Jean-Claude Ellena, we've seen it.' And I say, 'Yes, you're right. I don't copy other people. I do not copy from the market. *Je fais du Jean-Claude Ellena.*'" I do Jean-Claude Ellena. "They put them into the mass spec, and when they discover that I did it with only twenty ingredients, they say '*Mais putain! Tellement peu d'ingrédients!*'" Holy shit! So few ingredients!

Then there were the ways in which he was just plain cynical. One thing Ellena viewed with a decidedly jaundiced eye was the

American obsession with synthetic musk molecules. These had a rather strange olfactory backstory.

Musk—real musk—is a molecule: (3R)-methylcyclopentadecanone, but its popular name is muscone. (The chemical name/common name pairing is quite common; there is, for example, the molecule 2-acetoxybenzoic acid which also has a popular name: aspirin.) Muscone is found, at a concentration between 0.5 percent and 2 percent, in something called Tonquin musk, a richly stinking secretion mixing hundreds of molecules that comes from a gland inside the male musk deer. Extract the rich secretion, separate out the little bit of muscone from it, and by itself this one molecule has the warm, sensual, rich scent of clean warm skin. It's only this molecule (not the full, richly stinking, animalic Tonquin musk secretion) that perfumers call the scent of musk.

The problem was that you had to kill the musk deer to get its gland. This made it harder to find (it also became increasingly illegal for ecological reasons). What to do. In 1888 a man named Albert Bauer was searching for new explosives and noticed that when you did a chemical reaction with TNT (trinitrotoluene) and tert-butyl halides, the two produced a new synthetic molecule that smelled almost exactly like (3R)-methylcyclopentadecanone—i.e., like musk. The molecule, found via this explosive that had nitros in it and itself containing a nitro group, was called a nitro-musk. Named Musk Bauer, it became the first synthetic musk used in perfumery, the replacement for the natural material.

Chemists then found other synthetic musks: Musk Ketone was created in 1894—a lot was put into *Chanel No. 5, Brut,* and many more perfumes—and Musk Xylol, Musk Tibetene, Musk Ambrette, and Moskene. The problem with the nitro-musks is that they had NO_2 groups that were unstable and degraded into different colored products. (Also some turned out to be toxic and were banned.) So the industry set out to find replacements for the replacements, and they found the polycyclic musks. These—Galaxolide was a famous one—had no NO_2 groups. But the polycyclics, it turned out, didn't biodegrade, so they went back and looked around in the natural musks (called macrocyclics) and modified them to come up with new synthetic musks like Habanolide and Musk R1. And then the linear musks, like Helvetolide and Serenolide, which were cheaper and had all the advantages of macrocyclics, plus in some cases also slightly fruity twists.

The strange thing was, says Ellena, what Americans used them for. Essentially these molecules were used as perfumes for laundry detergents. "The first scented detergents arrived in France from the U.S.," said Ellena, "where they existed since before the war." He listed them: Colgate, Procter, Unilever. "They created these molecules in a lab and smelled them, and they smelled strong, and they were quite stable. They were also cheap to manufacture. They weren't water soluble. And these molecules smelled good in quote marks, which is to say people like them, which is certainly one definition of good. So how to use them. They found out that when they put them in detergent they stuck very well to fabric during the wash—again, water insoluble—so they put lots of them in cleaning products and put these products on the market. What happened? People associated this smell in their detergent with the idea of 'clean.' The molecules rubbed off from their clothes to their skin, and then people become used to having these molecules on their skin, so they came to associate wearing these synthetic molecules with smelling of fresh laundry, and so today when people smell these synthetics they say, 'Clean' and they say, 'Skin,' and they say, 'It smells of Me.' " He added with a smile, "All of this unconsciously, of course."

(It keeps perfumers on their toes. You have, for example, a truly first-rate scent by l'Artisan Parfumeur, *Mûre et Musc*—Blackberries and Musk—one of the more wonderful perfumes around, expertly crafted, and it uses these synthetic musks, which means that if the perfumer doesn't dose them right, it could smell a little bit like Tide because it smells like the musk

1,3,4,6,7,8-hexahydro-4,6,6,7,8,8-hexamethyl-cyclopenta-gamma-2-benzopyran, which is sold under the trade name Galaxolide, i.e., the smell of laundry detergent. So you need to be careful.)

Ellena evinces disdain for certain materials, but it is never a prejudice against a category and it is never for ideological reasons. It is always an aesthetic choice. "In the 1990s, I created *Déclaration* for Cartier," he says, "and I used synthetic musks. Today I don't use synthetic musks, not because I think they smell bad—I don't—but because I think they're easy. They're a crutch. When you don't know how to sweeten something, you put in sugar syrup. It's very easy. But I use them when I want them." He names *Angéliques sous la pluie,* a gorgeous floral perfume he did for Frédéric Malle's collection. "Sugar is easy to use," he says and makes a dismissive motion: "Sugar for sugar's sake: *no.*" Ellena loathes dihydromyrcenol as much as I do. It's a molecule that smells like sink cleanser spilled on an aluminum counter. (Actually I've met few perfumers who don't dislike it—"I'm at war with this molecule!" one of them e-mailed me when I brought it up with him—though often not because of the way it smells but simply because aesthetically it has become a huge eye-roll of a cliché.) Ellena identifies it (not enthusiastically) as opening what he calls "perfume's *phase hygiénique.*" (He simply noted, "Dihydromyrcenol was hugely used to scent laundry detergents," and stopped there.) To Ellena, you could pinpoint the beginning of the hygienic trend as *Paco Rabanne Pour Homme* (1973), which was leading away from leathers and ferns and spices and straight into the fluorescent-lit drugstore detergent aisle, but the dihydromyrcenol revolution was formally launched in 1982 by Pierre Wargnye, who exploded it onto the scene by—ingeniously at the time—making it a jaw-dropping 10 percent of the formula of *Drakkar Noir.* There'd never been anything like it, and frankly it was kind of sexy to smell like your shirts just out of the dryer. It reached its ultimate apotheosis in 1988's *Cool Water,* a dihydromyrcenol orgy made by Pierre Bourdon, who dumped in 20

percent. Six years later, perfumers Alberto Morillas and Harry Frémont cast it in a starring role at 6 percent in their *cK One* formula,
and even then one could argue that the *cK One* innovation was both
significant and authentically creative. Ellena's view was that these
were valid and even creative perfumes in their own time. Now that
that molecule has been dumped into 8 million masculine scents, it's
a boring cliché, but personally I dislike natural lavender for the
same reason: natural or synthetic, a cliché is a cliché, and today no
one should ever touch the stuff.

Ellena also categorically refuses to use Galaxolide, this synthetic clean smell. He is, says his daughter, "tyrannical" about the
laundry. He has asked Susannah to re-wash sheets to remove the
synthetic clean molecules because he couldn't sleep, and every new
detergent that comes into the house he must smell and approve. In
hotels, he covers the pillowcases with his shirt or a sweater to coun-
teract the scent.

And still he did not, on principle, oppose any new molecules.
Ellena agreed with Frédéric Malle, who observed that technology
was playing its usual role in art. The new synthetics, for example,
and the scents they allowed you to create that you never could have
before. "They're more *performant*," said Malle, "and can be overlaid
in a much cleaner way, without interfering with one another."

The most daring perfumers were all groping for new modes of
expression. Some were developing a kind of transubstantiation; they'd
start with a raw material and create a final perfume out of it that rendered it into almost metaphysical form. The niche Italian house L'Eau
d'Italie hired the perfumer Bertrand Duchaufour, a master of olfactory chiaroscuro, to produce the mesmerizing, shadowy *Bois d'Ombrie*,
and putting it on was like slipping on an organza silk shirt. You saw
the material—the silken, shadowy wood smell—it had immense, gorgeous form, and yet it was transparent, if not invisible, and you saw
every bit of skin underneath. *Bois d'Ombrie* was an olfactory garment.
It sat on you like the most gorgeous Givenchy haute couture piece,

and yet it became part of you somehow. The ingenious 2 by Comme des Garçons was the smell of a person, heightened into the surreal.

Perfumers were pushing the limits of the art form. "Years ago, I would be working on things," said Malle, "and [the great perfumer] Jean Amic would smell them and say, 'It's a wonderful smell. But it's not a perfume.' It took me a long time to figure that out, but it's the key to my own collection. Sometimes a thing is good in a room, but it won't become part of you." Malle observed, "The great aspect of *Eau Sauvage* is that while people think it's an eau de cologne with a flowery green twist to it, what actually makes it work is a leather scent that links the freshness to human skin."

In crafting *Nil*, Ellena was embracing technology—the new materials, the new techniques—but not (he would underline this forcefully) for their own sake. When he thought about what he was doing out loud, Ellena said abstractly intellectual things like, "Perfume is an adjunctive sense, and time is indissociable from its creation. Time is also a sensual element, a sort of action at a distance which inscribes itself in memory." (Then he stopped, looked ironically amused, and said, "That was very French, wasn't it.") At other times he could say concrete things: "I like the simplicity of objects that reveals their true beauty. A watch that's too complicated won't work for me." He wore a Mirus watch he'd come across years ago by chance at an antiques dealer during a walk in the town of Mâcon. It was simple, rectangular, pure. It had been on his wrist when he'd created *Un Jardin en Méditerranée* for Hermès. Since then, he'd been unfaithful to the Mirus. He'd changed to an Hermès Arceau Équestre. He liked its graphical simplicity and the large numbers, for his myopia.

Although he did not say it (it wouldn't have been his style, that sort of self-revelation), like all of us Ellena was looking for his place in this new world. In one of his notebooks he had written down something from the poet René Char. "*Ce qui vient au monde pour ne rien troubler, ne mérite ni égard ni patience.*" What comes into the world to disturb nothing merits neither attention nor patience. "It's violent, this sentence," said Ellena. With each iteration of each perfume he made, including with *Nil*, he was altering his world, disturbing the order, and he was also being altered (and risking falling flat on his face), and there was a certain violence in this process of artistic creation. But he agreed with Char. If you couldn't do it, or were afraid to, stay home.

In his search for Hermèsness, Ellena naturally observed Jean-Louis Dumas. Dumas headed a large, specific tribe—the family—and they were organized in the at-once completely fluid chaos and absolutely ironclad order of the wealthy French familial social order. Dumas had a strong touch that age had made both more velvet and more assured, but whatever he ran he ran like clockwork. To journalists, he would say gnomic, deeply French things about Hermès perfumes like, "The Hermès perfume aesthetic is the idea and the simplicity. The idea is fundamental. It is like a drop of water on a leaf after the rain, a sentiment expressed in the eye of the other, a miniuniverse that carries the force of all it has created." He would also say practical, less-French things like, "Roudnitska's *Eau D'Hermès* is a school of realism. The smell of the leather saddle." (Ellena thought about leather scents.) "We are saddlers. Even in our perfume, we have to make the leather saddle elegant and useful."

CHAPTER 9

THE WRITER TAD Friend nailed it in his "Notes on the Death of the Celebrity Profile." He opened this way.

Exhibit A. A few years ago I had a brief and mutually unsatisfactory conversation with one Charlotte Parker, Arnold Schwarzenegger's PR flack at the time. I had just met Arnold while reporting a Planet Hollywood story for *Esquire* when Parker drew me aside: "We'd like to get Arnold on *Esquire*'s cover," she said. "He's been everywhere else."

"Oh?"

"How much time with him would you need?" Parker asked. "Half an hour? Forty-five minutes?"

"Well, *if* I were going to do it, I'd want to hang around over a few days," I said, rather stiffly.

"Oh, God," she said disgustedly, "this wouldn't be one of those profiles where you try to *figure him out*, would it?"

Exhibit B. *Us* magazine, December 1997.

Us: "What [do] you loathe about yourself?"

Richard Gere: [Laughs darkly] "You are someone I met ten minutes ago, and now you want to get into the deep, dark questions about my being?"

Friend more or less pinpoints the way I felt going in. The conventions and clichés of standard fashion journalism (perfume is treated as fashion; I don't make the rules) are toxic. Reporting on scent is akin to a rocket's launching: You expend your greatest energy fighting free of the public relations muck. Once that's done, if you actually find any substance, the real work is (comparatively) weightless.

With Sarah Jessica Parker, the process was compounded, it seemed to me, by the question of how the hell one was supposed to get anything real out of a celebrity. To make it worse, Belinda Arnold had sent me a press release. Perfume industry press releases are like the dark matter of the universe. No one can actually see them; no one reads them; they are recycled boilerplate vanishing even as they appear. Each one is built out of plastic predigested identical components on a fixed formula: The first sentence breathlessly announces the magnificent news, the next two give you "the wider picture," the PR person writes quotes that she attributes to the corporate people, who OK them on their BlackBerries in airports, the thing is poured into a mold to set and, once dry, sprayed into the in-boxes and fax machines of a thousand journalists and editors, who dump them in their trash cans. (The exception being when the editor, lazy and/or craven, simply runs the release. This happens all the time.)

This press release was standard.

Contact:
Belinda Arnold
Lancaster Group US
Public Relations Director
212.389.7428
E-mail: belinda_arnold@cotyinc.com

COTY INC. PARTNERS WITH SARAH JESSICA PARKER TO DEVELOP AND MARKET FRAGRANCES
Agreement Marks the Actress' Debut into the Beauty Industry

NEW YORK, February 8, 2005—Coty Inc., a leader in the beauty industry, announced today the signing of a global licensing agreement with Sarah Jessica Parker to develop and market a line of fragrances . . .

And so on. The partnership was the "first entrance" by the internationally known actress into fragrance. It would be designed to reflect the talent, style, and confidence "of which she has become a symbol."

The fake quotes were carefully predictable. In PR, the star is always thrilled, and so that was what they had her say. " 'The creative energy and dedication of the Lancaster Group is exactly what I am looking for in a beauty partner,' said Sarah Jessica Parker. 'Their admiration and passion for their brands is what I was impressed by, and I look forward to working closely with [them] to bring my first fragrance to the market.' " The standard manufactured quote from Coty's CEO Bernd Beetz (Parker's "qualities and values make her a global icon") and then, aiming squarely at Wall Street and the stock analysts, a plug for Coty, how the new partnership was "yet another example of how we are able to succeed and build stellar brands . . . by moving faster, thinking more freely, and taking our ideas further."

The quote they made up for Walsh was, naturally, designed to be more intimate. " 'We are delighted to add Sarah Jessica to our portfolio,' said Catherine Walsh, SVP Marketing, Cosmetics and

American Licenses." There was the obligatory application of the words "iconic" and "excitement." The press release is, quite literally, a form of poetry with its own imagery, conventions, and tropes. It is also an extremely professional execution of corporate policy. I once angered Carlos Timiraos by writing something in *The New Yorker* about a soulless Kenneth Cole perfume, and Timiraos's e-mail said, "I'm very upset about what you said about *Kenneth Cole Black.*" Not *Black*—it was *Kenneth Cole Black.* He carefully wrote out the full name of the perfume. These guys are always respectful of the designer, always careful with the brand, always on message.

The question is the degree to which you're ever going to get something real. It wasn't, incidentally, all that different with Ellena. He had boundaries and expectations, ego and moods, and we dealt with those in the usual way during the process. The logistics of covering the two of them were different, but the activity itself, ultimately, was not.

I had dinner one evening at Josie's on Third Avenue with Melinda Relyea, Parker's assistant, because we found we had a friend in common, Clay Floren, so the three of us got together. It wasn't an interview about Parker, and I wasn't taking notes, but Relyea turned out to be an interested observer of her situation. She commented at one point—she was speaking about her job—"No matter who we're with, movie stars, producers, fans, whoever it is, Sarah Jessica never introduces me as her assistant. She says, 'This is Melinda Relyea.'" It was significant to her, and she said so. It wasn't, she said, that they never had tense moments. They did. "But at the end of the day it's me and her in the airplane seats next to each other, going over whatever we need to go over, and she's never, ever not made that work."

I was, at every point, conscious of the care with which Parker managed her fame. I was riding my bike up Sixth Avenue after our first day spent together and turned right on Thirty-eighth, and she called me just to ask how did I think it had gone and was it OK?

and wow, she was always so nervous about these things. I'm standing with my bike between my legs on West Thirty-eighth Street talking to her on my cell phone, and both of us know exactly what she's doing: She's taking care of *The New York Times*, and that's fine, she should be. The fact that she's being nice to me is analogous to any business transaction. It only feels stranger because the voice is so familiar, and because it's all handled in personal terms. I was talking with her one day about TV—I'd been approached by a producer who wanted to make a television series with me about perfume—and she gave me some advice, mentioning a director who was doing an interesting television project. A friend of hers, she said. She added, "Sam Mendes." I said something like, Sam Mendes, he's great! She agreed. And then after a moment she added, "And he has a very beautiful wife."

I actually only realized afterward that this was a tiny bit of management, the line she was supposed to say because she's a careful professional; whether she believes it is beside the point. I knew Mendes was married to Kate Winslet, and Parker knew, even if I didn't, that editors who edit the things written about her would ask the question: "Oh, she talked about Mendes? What did she say about Kate?" Parker had supplied the answer, one that conveyed the correct respect, was entirely appropriate, and of course revealed nothing about her. It made me feel funny, because I didn't want her to feel obligated to say those things to me, but that was just my being naive.

As Tad wrote, you feel this weird obligation to *figure her out*. And she's just not going to get into the deep, dark questions about her being, and yet at the same time I've had trenchant, indelible conversations with complete strangers on the 4/5 subway, people who've said things to me and I to them about those questions, things that I wouldn't say to anyone who knew who I was or who I'd ever see again. Parker and I liked each other as much as any two people who meet professionally and get along, and even as we were doing our jobs, she said real things to me and I to her. We had one real and very intense argument for about ten minutes at Mary's Fish Camp, a restaurant in the Village, and then consciously put it to rest. We talked, when we weren't talking about the project, about things that mattered to us. She wasn't going to waste that precious time if she could help it, and I wasn't either. Everyone tries to enjoy their work and live their lives well, and this is what you think as you go through.

With Ellena it was a bit different inasmuch as he is more obviously guarded and careful than she is. And oddly he has, in a way, an even more highly developed sense of the correct presentation of his persona as perfumer to the public. But he also put that aside at other moments and was free and open. The normal fluctuations. With both of them I talked about politics, very frankly and off the record, and that was quite interesting. With both I talked about their careers.

I'm sure this business of my watching and revealing a year of their lives during these complex projects was as strange for both of them as it was for me. Ellena had an entire house trusting him on this hugely public journalistic project, a house in which he was moreover the new guy, and he had to shoulder the whole thing, not screw it up, make it work.

And Parker: She's telling some guy she doesn't know a story about a business venture she's created, for which she's responsible, one that involves hundreds of other people, their jobs, and millions of dollars of other people's money, a project that uses *her life* and her tastes and her name, it's her job to represent it. And she's got to trust this guy to get it all right?

                                                        ≫◎

The second meeting at IFF takes place on January 13, 2006. They assemble again in the same conference room.

At this point, they're actually not sure if they're going to change the original *Lovely* scent in the new product. Walsh, Parker, and Timiraos have been toying with the idea of a new olfactory angle. Parker actually came up with the idea after the first meeting, describing it as "adding salt and pepper." What and how remains vague.

What they want to create, Catherine Walsh reiterates to everyone at the large table, is an unflanker. If they change the scent, she says, you'll still have to recognize *Lovely*. The question is how they might change it; that is, what does she like? (She glances toward Parker.)

"So today," says Yvette Ross, the IFF account executive, taking the conversational ball smoothly from Walsh and moving into the session, "is about finding out what Sarah Jessica Parker likes."

Parker opens her eyes wide and smiles. "Well!" she says, "*that's* fun."

Ross is handing out sheets of paper with olfactive categories printed on them. She's going to be presenting scents to Parker, the project's creative director, and getting her feedback. This is crucial to any perfume creation process in which the creative director's vision, not the perfumer's, rules, which is the case in 99 percent of perfumes on the market. (This is why Ellena's creative role at Hermès as perfumer is so highly unusual. While Dubrule and Gautier are officially cocreative directors of *Un Jardin sur le Nil* and possess—if it comes to it—veto power, he operates both as perfumer *and* a first-among-equals creative director of his own perfumes.)

As Ross organizes her scents in the IFF conference room, the players are also conscious that there is an ulterior motive for this meeting, which is that whether or not they create a different version of *Lovely* for the new product, Coty and IFF both assume the SJP franchise will be launching another completely new Sarah Jessica Parker perfume. So today's meeting will serve double duty, a creative meeting both for the *Lovely* product they're working on today and for that future scent they all hope to start on soon.

Le Guernec and Gavarry have spent months working on scents to show Parker. They've created a few dozen, each a sort of olfactory diagnostic test, designed not so much to please her—or necessarily to become actual perfumes—but rather to give them calibrated soundings of her taste in smell.

The IFF marketers have grouped these into three categories. First is the somewhat cheesily named "Sarah's Garden," all floral scents; they're perfectly aware Parker generally hates florals but want to run them by her; they also need to know what she dislikes. The next category is "Background Notes: Dark Animalic Incense" (which generally she loves) and then "Skin and Musks." The clear glass vials of liquids are lined up in precise, neat rows, the stack of clean white paper *touches* like cardboard ammunition.

Ross begins loading, dipping the *touches,* but Parker is frowning, looking over the scents to come. She is as conscious of her primacy in creating the next perfume as she is of her lack of training. Each elixir represents a technical feat, a blend of molecules, that she herself would be incapable of creating. The two experts are before her, and she asks Le Guernec and Gavarry, "I'm curious"—she hesitates—"isn't it important that these be notes that *you* like?" She glances at Walsh, who is watching her carefully. "Because can't that somehow—you know, help you make it great?"

Le Guernec glances at Gavarry. He looks back at Parker. "We'll make it great."

They're here to realize her vision, that's their job, but it is a question that is virtually never asked of perfumers, and Le Guernec is authentically surprised, and touched. It goes to the heart of the debate among perfumers—when they talk to each other out of earshot of their clients—from the moment they enter the perfumery school to the moment of retirement: Are they artists with their own visions and aesthetics producing art, or are they hired artisans whose job is to be simply the amanuensis of the creatives and the marketing people?

Ross hands out heavy, gleaming new metal blotter clips that hold the slips of paper conveniently upright ("Wow," says Parker, "you're *giving* us those? You guys must be rolling in it!"), and the first of the battery of smells is launched. Its name is Sexy Potion. Gavarry created it.

Everyone smells it. Silence for a moment. "OK," says Parker, "here's what surprises me. It's more powdery than something I'd have thought I'd like. I mean a 1940s talc. Which is strangely appealing. It's literally milky. High dairy. For those lactose intolerant, this is not a fragrance for them. It's cool. For what we're going for, it would be good to add some complexity?"

She stops, looks at Ross. Ross has bent the tip of her blotter—the part she dipped—up. "Why did you bend it?"

Ross: "To make sure it doesn't touch the table."

Ross hands out Erotic Skin.

Parker: "It has the depth of great ambers. Erotic Skin. It's *definitely* skin. I like the warmth of it. It's not rough or unpretty." She carefully bends up the blotter's tip and places it so the tip doesn't touch the table.

Walsh: "There should be a trend to go back to amber."

Parker: "There's an amber I love, which is Prada, it's terrific—"

Le Guernec is electrified. "We did that . . . !" he says. By "we" he means IFF, and he points—it's very sweet, he's like a schoolboy, his finger moving eagerly—at Gavarry. It was Gavarry, along with two other IFF perfumers, his father Max and Carlos Benjamin, who created the Prada perfume.

Parker is surprised. "Wow! *Great!* Seriously! Well, what *I* think is brave about Prada is that it has sweet in a strange way, and it has so much depth. You feel stone. It's not some feminine thing." She really loathes anything typically feminine.

Nicolas Mirzayantz, IFF's senior vice president for fine fragrances, an extremely tall and dynamic man with his trademark gigantic muttonchop sideburns, comes in and congratulates Parker on her success. "And you're so much part of it!" he says.

"For better or worse!" says Parker.

They talk kids; their children go to the same school in Chelsea.

Walsh is still smelling the Erotic Skin. "I don't know if this added to *Lovely* would get us where we want to go, but this is definitely where the market is going."

Next is Elixir: Dark Amber. It's powerful and shadowy with strange dark angles.

"Whoa!" says Parker. "Unbelievable!" She loves this darkness. She can't get enough of it. Walsh pretends to panic; this is a resolutely noncommercial scent.

Ross (placatingly): "It's just an indication of how far you can go."

Walsh: "If Carlos were here"—Some heads turn, realizing Timiraos is not there; "He's on vacation," says Walsh—"he'd be kicking me under the table."

"There's real dry, musty attic in here," I say.

Parker: "Yep."

Walsh: "Damp."

Parker: "Yes."

Walsh: "This is so much more real than Sexy Potion."

Parker: "This is *so* much more sophisticated." Smells it, sighs, puts it down with regret. "That's so neato."

Ross: "Let's take a quick break. The next ones are florals."

Parker (with a smile, but firmly): "Which I tend to respond less well to."

Ross (with a smile): "Uh huh." Ross experienced Parker's dislike of florals firsthand during the process of creating *Lovely*. "You said you liked Plum Blossom, so Laurent made one up."

Parker's down with that. She says, "Great!" She jumps up to get a water, turns, asks, "Anyone want anything?"

She comes back. Seeking to be amenable, she says, "You must have a hint of floral for commercial success."

Le Guernec encourages this: "Everyone has a floral piece."

Gavarry: "Even Prada has a floral."

Parker (she's not at all convinced but is trying): "Even Prada!"

THE PERFECT SCENT

Ross deals out scents as the perfumers track Parker's expressions and movements like watchmen in a tower. This first accord is called White Bells.

Parker: "Whoa. Total lily of the valley. It gets a little"—(she's smelling it)—"chaste for me. Churchy."

Le Guernec: "Hm." He leans into it again, analyzing it in short, shallow inhalations.

Parker: "Righteous." Looks at him. "Moralistic."

Le Guernec clearly decides it's time to face facts: "So you're *not* a floral person!"

Parker: "My mother loves lily of the valley, so I carried them in my wedding." She tries to look a little apologetic for not being a floral person, but really she just isn't.

"Centifolia Rose," says Le Guernec. The next accord. "Rose is difficult," he adds, "because everyone goes back to face powder or toilet paper."

Parker: "Yes!" She laughs, and Mirzayantz grins and says to her, "Don't blush."

"Red Velvet Rose"—this one smells amazing. Parker notes the soap angle: "Smokey . . . slightly waxy . . . amazing! It's got depth; it's got tactile properties." It contains an LMR rose material and Mirzayantz tells her about LMR, how Monique Rémy fought for the perfumer.

Parker starts picking up the *touches* and putting them together, trying combinations. She talks about how they might use them, proposes things. Le Guernec hands her "Plum Blossom." (When she went on *Oprah*, she did a produced video segment walking around New York, showing the things she loves: a tchotchke store in Chinatown with silk pajamas, a burger place on Fifth Avenue in the Fifties that Matthew goes to for chili burgers, a painting in MOMA with plum blossoms.)

Gavarry: "It's like a combination of fruity and nuanced notes with flora. Creaminess."

Walsh: "This as a steady diet would be cloying."

Parker shoots her a grateful look. "I get soapy."

Gavarry: "Sarah Jessica, when you say 'soapy,' is that negative?"

Parker's face says yeah, negative. "That soap," she says, "that sits forever and ever in the guest bathroom that matches the color of the towels and then gets dust on it? It just feels . . ." She makes a desperate look.

Gavarry nods. Got it. Fine.

Parker talks about the notes in Lauder's *Aliage*, the green notes, her dad's pipe. Ross starts sending out the "Dark Animalic Incense" category, and so she talks about Marlene Dietrich and how she wore *Creed*.

"Gold Incense."

Parker smells: "Wow! I love that," she says. "It's cold and warm together." Smells it again. "It's *addictive*."

Ross and the perfumers share a look. "We figured this would be Sarah Jessica," says Ross.

"Hearts of Fire." Walsh hates it. She writes "Nauseous" on her paper. Parker agrees. "Not as brave." She goes back to "Gold Incense."

"Amber Crystals." Creamy and earthy. She likes it OK.

"Raw Amber." Parker picks it up and is completely still. Then suddenly grabs Ross's arm and says, "I freakin' love this." She doesn't emphasize the words, simply states them. She sits there, inhaling, moving. A lock of hair falls over her face. She murmurs, "Sweet."

She smells the geranium.

Le Guernec: "Geranium has a lot of mint. A lot of coolness."

She says to him, "Gorgeous." She blinks. "Is there a perfume called *Gorgeous*?"

Yes, says someone.

Parker shrugs. Oh well. She adds, "Very Barbra Streisand."

The last scent is "Ultra Suede." She loves it. "This is *gorgeous*. This is *gorgeous*. I *love* this!" She says "creamy leather" and "buttery" and "soft." She squints. Identifies vetiver.

You can see Le Guernec's eyes open. "Hm!" he says. "She's good."

Walsh: "I don't see it mixing with *Lovely*."

Parker: "Neither do I. But we'll save it. As [the stylist] Pat Fields says"—lowers her voice, belts out—"'*Don't burn it!*'"

The next accord: "Southern Magnolia." Parker is . . . indifferent. Doesn't hate it.

She goes back up the *touches,* smelling each one rapidly. "My gay friends would *love* for Gold Incense to be part of it. It's so natural, earthy."

Mirzayantz: "It's not until the twentieth century that perfumes were gendered. Before, they were for everyone."

Parker and Walsh, facing each other across the table, start to do their triage: "Gold Incense," "Raw Amber," "Ultra Suede," "Elixir," "Red Velvet Rose." Parker: "Perhaps we could use 'Gold Incense' in our *Lovely* skin product and, if I'm lucky enough to do a second fragrance, separate from *Lovely,* reintroduce it, in a completely new idea, as a note."

They'll keep "Southern Magnolia" in the game for now. "Centifolia" is just "too high." "Amber Crystals," "too light." They get cut.

I suggest they show her *Cabochard. That's* dirty. They go and, after a few minutes, retrieve a sample of this 1959 perfume and give it to her. "Iva Rifkin! This is my friend Iva Rifkin! This is unbelievable. *So* sophisticated. Iva Rifkin!" She's enraptured, then instantly despondent. "But if girls are buying Britney . . ."

Walsh: "My dear, you are already breaking the mold."

They talk perfumes: *Aliage* (Parker: "I wore it!") and *Cinnabar* (Walsh, guiltily: "*I* wore that.").

꙳

Everyone goes down to the IFF perfumery school to see Ron Winnegrad, who heads it. "Your son's favorite painter is van Gogh, right?" he says to Parker. He gives her a children's book for him, *You Are My I Love You*. "Perfect," she says. "Every day I tell him I love him, and every day he says, 'But *why* do you love me? *Why?*'"

Winnegrad is talking about synesthesia, a neurological condition in which senses cross-mix inputs—hearing colors or smelling music. "You're born with it, but you can develop it too. So we're going to try to cultivate a little synesthesia today." He makes Parker read a collection of words that are colored, but she has to—as fast as she can—say the color of the word and not actually read the word. It's incredibly difficult. He shows her how evolution has programmed the nostrils to take turns constricting from right to left, the open one allowing the lighter molecules to rush through and hit the olfactory bulb, the constricted one trapping the light molecules and letting through only the heavy ones.

Winnegrad has her smell the original *Ralph Lauren*, which is a brilliantly constructed perfume, an extremely and wonderfully odd combination of fresh plant and wool gabardine and clear wood, like opening a clean wardrobe to find a tropical fruit tree growing inside it. He has Parker write down adjectives, paint with watercolors.

⁂

They head to IFF's Technical Applications Lab, where TAG (the tech applications group) does its thing; someone says to the group in the hall, "You going to TAGland?" The lab, large, spotlessly clean, and formica-white and looking just like the spaceship in *2001: A Space Odyssey*, is where they do things like testing bases and choosing colors for juices.

Steve Semoff is there to meet them. "I'm on again," he observes dryly.

"Take it away!" says Parker. She turns and sees the robot and stops dead. One wall of TAGland is covered by a huge robot that mixes scents. It is a giant metal machine with dozens of tubes rotating around its middle. "It's a robot," Semoff says, and explains to her how it mixes materials, milk shakes of allyl amyl glycolate and undecalactone and ethyl vanillin.

She is amazed by the level of technology necessary to produce a sample. The robot's name, says Semoff, is Elvis because whenever people asked, "Is the robot here?" the response became inevitably "Elvis is in the house."

Semoff gestures at five products sitting neatly on a lab table. "Now we're going to look at the market," he says, and decide which of these might serve as a good model for Parker's product. (Looking at the competition is—as it is in every industry—standard practice. One executive told me, "Your first thousand SKUs you sell to your competitors; clients only start buying them at the thousand-and-first.")

Each of Semoff's five products is a perfume in some kind of skin gel. First is a *Chanel No. 5* gel, which Chanel calls Elixir Sensuel. The product is, quite simply, *Chanel No. 5* perfume in hydroxy propyl methyl cellulose. The second product is another perfume in an alcoholic gel—Parker tries each one on in turn—the third a body oil. Parker particularly likes the fourth, a silicone-alcohol blend. The fifth is a body milk, but a body milk you can spray. Parker goes back down the line. "Number four—could you spray it?" she asks. Semoff replies that some silicones can't be sprayed because there's an inhalation issue.

They examine them. Walsh and Parker agree that, as Walsh puts it, the milk "takes you to an ancillary product and not a fine fragrance." So that's out. Parker doesn't particularly like the oil, so that's out. The gel she finds too sticky.

Number four wins. It's a liquid silicone lotion, a blend of dimethecone, an expensive material. "What I like about this," says Parker, "is that it gives you a silken feel, but it's dry. I like number five, the milk, as an ancillary."

"You can load the silicone lotion up with all sorts of stuff," observes Semoff. "Vitamins."

Parker: "Yeah, yeah, whatever, baloney." She's impatient with the gadgets, with tricking out a product.

Semoff, who completely agrees, says dryly, "Yeah, whatever you want" and peers around the room.

Parker's got to go. Two marketing people, very hesitantly, signal a need to talk to her, one last thing. What's up? says Parker.

They're a bit nervous. It's, uh, well, they need to ask her to do, uh, a press thing?

Parker looks at them for a second. "What is it?"

Like, a list of questions? They could e-mail them to her?

She's so fine with it she looks confused. "Sure!" she says—why are they bothering with all the diffidence?—and then with equal measures of resolute professionalism and resignation, she gives them the official answer: "Tell them I'm completely amenable to it."

Well (they look at each other), there was apparently some— issue? With one very protective publicist working for Parker?

Ah . . . She's quiet for an instant, clears her throat in a way that means she'll take care of it. She finishes with "I'll do whatever it is that you need me to do."

She says her good-byes, kisses, hugs. Says to Semoff, "We hardly saw you! You come in right at the end!"

"I'm the closer," he observes dourly.

She gives him a confidential, New York gangster look, lowered voice: "Yah uh excuhllen closuh."

⁓◉

I met Peter Hess, SJP's agent at CAA, at a black-tie event. The usual gauntlet of photographers was corralled in a vast squirming, tentacled rectangle beside the red carpet, aiming its hundreds of recording devices, an electrical firing squad. The two of them moved down it, Hess in tux, Parker in gown. Hess escorted her, gauging the rhythm, supporting her glide, a Duke of Edinburgh with an expert hand lightly on the controls. He's a handsome guy, compact and smooth and controlled, sort of the prototypal agent. They could cast him. The cameras tried to suck her into their lenses, and Hess was collaterally strafed by the flashes cloaking her in white hot light. I noticed that he was impervious, as was she, to the photon bombardment; not a single blink, no matter the intensity of the barrage.

I was standing just inside, watching, and she walked over and gave me a kiss and a hug and exclaimed, "Chandler *Burr!*" and, gesturing around her, "Isn't this *something?*" and we chatted about it. She turned to greet someone, and he put out a hand and said, "I'm Peter Hess." I said, Oh, *you're* Peter, and started to introduce myself and he laughed and said, "I know, I read your piece."

Yeah?

"So, like, uh," he says, wryly, "I love that she comes to me with this idea for a perfume, and I'm in *The New York Times* as being like, 'Whatever, I'll get back to you.'"

It wasn't like that?

"Well, *no*," he said, and hesitated. More serious now. He had the agent's instinct against putting himself into his client's story and the agent's desire not to be left out totally. "I mean, look, I was dedicated to it."

She told me that. Seriously, she made that clear. I grinned. I didn't mean to dis you, I said.

He shrugged; he really just found it funny. I was trying to remember what the hell I'd written. I went back and looked at the piece later, and he was right; she'd presented him to me in, frankly, pretty broad strokes, and the part in between her saying, "I went to my agent and told him my dream of doing this thing" and the part where she said, "I got a call from my agent and it all started happening" she hadn't exactly filled in. It did make him sound like a real impresario—"I made it happen, baby, and you're on in *five*"—but I can see how a guy would want his character fleshed out.

It in fact took about half a year for me to find out just how big a role Hess had played. There is always the story of making the story, and that was him. And the Parker/*Lovely* backstory is that here was this huge perfume success, and it almost never happened at all. No one would take her. Hess couldn't find a licensee that would sign a perfume deal with her. He didn't tell me that, of course, nor did anyone at Coty. The industry did, and only circuitously.

Six months after *Lovely* had launched, I had lunch on Fifty-seventh Street with a senior exec at one of the biggest perfume licensees and mentioned Parker's excellent NPDs (her high ranking on the List). Slicing into his steak with a serrated knife, he said, "Yep," and then added evenly, "We turned her down." (I had just heard this from another executive.) He shrugged, good-naturedly. It wasn't bitter, it wasn't even regretful; it was simply a fact. I liked him for just stating it. No bluff, no excuses. The best of them are like that; hey, no one bats a thousand.

He added calmly, "Everyone turned her down." Everyone did. Hess had gone to all of them, L'Oréal, LVMH, Estée Lauder, Interparfums, etc., and said, "My client would like to create a perfume." Walsh told me very bluntly that of the singers, agents, movie stars, managers, and other celebrities and celebrity handlers who approach her, she turns down more than 99 percent. When the project sounds interesting, she and the other licensees run the numbers. And everyone had gotten to work and run them on Parker. One after another, they'd said no. In their views, the numbers didn't work: Their research told them that a Sarah Jessica Parker perfume wouldn't sell, or to express it in a slightly more sophisticated manner, that the perfume's potential sales didn't provide them sufficient assurance of profit given their projected costs for development and launch.

The Coty team—Walsh and Timiraos and CEO Bernd Beetz—did the same research. "At an early stage, Catherine hired a research company to validate SJP's appeal outside the U.S.," Timiraos told me much later when I finally came and asked him about it, "so we did need some convincing. And we were absolutely convinced. The research company completely validated her appeal. People were relating to her in Germany and the UK and France, which is where they studied it." Coty had looked at her box office around the world, focus-grouped her, asked people if they'd buy a Sarah Jessica Parker perfume, then ran the numbers very carefully. "You have to make sure the star is global."

In the end, Hess came to an agreement with Walsh, and they signed her. Several friends of mine have made offhand, somewhat dismissive comments to the effect that of course Parker's perfume became a huge success, she's a celebrity, everything works for her. This is incorrect. Mel Brooks once made a comment to the effect that the instant *The Producers* became a huge multimillion-dollar Broadway hit, people started using the word "inevitable" about it. Oh, everyone knew it would be a success, a gigantic money machine. No, Brooks said emphatically, no one knew. No one. Least of all him.

These perfume deals are standard, and they aren't. Every agent (and, from behind that careful remove, every client) negotiates different terms. One huge star made a licensing deal that was seen as stupid by everyone involved, the stupidity being huge distributions to the star that, observers felt, suffocated the perfume's marketing budget. Getting too much can actually harm you. Other times the brand screws things up. "Jesus Christ," an executive said to me, "can you believe Beyoncé got two million dollars up front *and* a percent of sales, and they didn't even put her name on the thing!"

Out of the blue one day someone e-mailed me a copy of a document that purported to be the licensing contract for Paris Hilton's perfume. Parlux is the licensee that rolled the dice on her; it's one of the smaller players. I have no idea whether the contract is real or not, but it certainly looks real, and those in the industry assure me that in any case it is authentically representative. It starts out

This LICENSE AGREEMENT ("Agreement") is made and effective as of the 1st day of June, 2004, by and between PARIS HILTON ENTERTAINMENT INC., with offices at 250 North Canon Drive, 2nd Floor, Beverly Hills, CA 90210 ("Licensor"), and PARLUX FRAGRANCES, INC., a public Delaware corporation with offices at 3725 S.W. 30th Avenue, Ft. Lauderdale, Fl. 33312 ("Licensee") (together the "Parties").

There is the Whereas clause for the master license given the Licensee (that's Parlux) by the Licensor, "Ms. Paris Hilton, an individual with a mailing address of c/o Ms. Wendy White," who is granting access to Hilton's hard-earned fame. ("Whereas Licensee is engaged in the business of manufacturing, promoting and selling Articles and Licensor desires to obtain the services of Licensee in connection with the manufacture, promotion and sale of the Articles bearing the Licensed Mark," etc.) This is legalese for describing what they all—the Céline Dions, Anna Suis, and Calvin Kleins—desire to obtain. (Givenchy and Dior don't have licensing contracts simply because they're entirely LVMH-owned brands, fashion and everything else, and LVMH "manufactures, promotes, and sells" its own brands. Chanel and Hermès don't because their operations are in-house.)

Then the substance. The license starts June 1, 2004, and terminates on June 30, 2009, and the contract gives Parlux the option to renew for five years. The contract defines *Territory* (Parlux receives world rights including duty-free stores, ships, airplanes, military bases, the diplomatic missions "of every country of the world," and "the world-wide web") and *Articles* ("Men's and women's fragrances, body lotion, body creme, body mist, shower gel, massage oil, dusting powder, shave gel," et cetera.). Parlux also covers its bases for the future by reserving right of first refusal to extend into "cosmetics, skin care, and candles." It has filed trademark on "Paris Hilton" in the United States in International Class 3 for fragrances

("Serial No. 78/412749"), and it takes the right to use "Paris Hilton" on all packing materials, hydrogen dirigibles, "and similar media presently existing or that may exist in the future." Presumably this includes interstellar craft. The contract carefully defines *Net Sales* as the sales price at which Parlux bills customers for Articles (which means Hilton only gets a cut of wholesale) less things like returned product, uncollectible accounts, discounts, and shipping that Parlux has to pay on any invoice. Hilton gets nothing from the sale of discontinued merchandise as long as it's sold at a discount of 25 percent or more "and then only to the extent that the aggregate gross sales thereof in any contract year do not exceed fifteen percent (15%) of total gross sales," a clause her lawyers put in to make sure they don't screw her by discounting, and sales of samples.

The Parlux lawyers, on the other hand, got Hilton's approval to their signing "other additional licenses for other similar lines" (i.e., if Jessica Simpson decides to move her brand to Parlux, even if Hilton hates Jessica, she can't block it), and Hilton's lawyers stipulate that Parlux will not attack Hilton's claim to her own name, a very serious issue; at the height of his immense success, the designer Halston lost control of his name, and it destroyed him almost instantly.

Confidentiality: "The Parties acknowledge that all information relating to the business and operations of Licensor and Licensee which they learn or have learned during or prior to the term of this Agreement is confidential." Et cetera.

Parlux guarantees Hilton its "best efforts to sell the maximum quantity of Articles therein consistent with the high standards and prestige represented by the Licensed Mark" plus a "suitable" sales force. It assumes full responsibility for designing and producing everything "and shall bear all costs related thereto." This means Hilton is completely off the hook for everything. Parlux also guarantees, without apparent irony, "workmanship of the highest quality consistent with [Paris Hilton's] reputation, image and prestige, distributed and sold with packaging and sales promotion materials appropriate for such highest quality Products." To give itself a concrete model, Parlux agrees to make fragrances of "prestige and price similar to that of Calvin Klein"—which is about as mass market as you can go and be considered luxury—"Tommy Hilfiger"—ditto—"and Ralph Lauren (excluding Purple Label)."

A single clause defines Hilton's creative involvement. It reads: "Parlux shall submit to Hilton for her approval... two sets of the fragrance, scent, packaging and other material, designs, sketches, colors, tags, containers and labels [the 'Approval Package'], which approval shall not be unreasonably withheld." In other words: Zero.

The creative input of the designers and celebrities and pro athletes putting out scents has always been more a matter of a good story whipped up for a gullible public. Donna Karan and Giorgio Armani and a few others are reputed to at least smell the juices sold under their names—no one knows if this is true; I've never seen them do it. Lev Glazman of Fresh lays claim to actually going into the lab and sitting with the perfumers at their benches while they build his scents, and he may, but again he hasn't done it in my presence. As a general rule, I suspect the claim that (fill in a famous name) was "deeply involved" is bullshit. From what I know, the greatest input most of them have is negotiating their percentage. One of the most successful designers in the world is actually rumored to be anosmic—incapable of smelling anything at all. Like members of a cult, his PR people dutifully recite how he "loved" this or that scent. PR people like to say the designers are "obsessed with" and "in love with" and "totally, like, crazy about" each launch, which makes them sound like deranged serial monogamists.

As far as I can tell, Sarah Jessica's involvement was extraordinary. This, of course, is the reason Coty allowed me inside.

If the contract is real, Hilton managed to get from Parlux something that only the stronger—or at least hotter—names command, a Guaranteed Minimum Royalty, paid to Hilton monthly on the first of every month; even if Parlux was unable to sell a single bottle, and before any royalties, it would pay Hilton. The contract, thinking forward, also stipulates the Minimum for each year if Parlux extends the license to 2014. At least the Parlux lawyers were able to get the Minimum credited against sales royalties; those will be paid on a quarterly basis within forty-five days after the close of the prior quarter's sales, plus any Guaranteed Minimum Royalty due. Generally designers and celebrities—licensors—are paid between 2 and 7 percent of the total value of the deal. If Hilton's perfume makes $50 million a year and she's got a 5 percent deal, she makes $2.5 million.

The contract stipulates that payments of the royalties and the Minimum are divided: 20 percent go every month to one of Hilton's lawyers, Robert L. Tucker, Esq., of Tucker & Latifi on East Eighty-fourth Street, and the rest to Paris Hilton Entertainment Inc. on North Canon Drive in Beverly Hills 90210.

Parlux also has to spend a minimum specific amount (it's calculated based on Net Sales) for consumer advertising: newspapers, magazines, billboards, direct mail, blow-ins, and billing inserts "(both scented and unscented)," product samples, window and counter displays, co-ops, et cetera. Plus a $3-million minimum liability insurance to cover Hilton.

The contract tangibly defines Hilton's work exclusively in PR terms. There is no mention at all in the document of her having any say in, or giving ideas for, or smelling any of the iterations of, or being creatively involved in any way with the smell of, the perfume.

> Licensor undertakes at Licensee's request to make Ms. Paris Hilton ("PH") available at reasonable intervals and for reasonable periods (which shall involve a maximum of seven (7) appearances during the first contract year and a maximum of four (4) appearances each contract year thereafter) for promotional tie-ins serving to associate PH with the Articles. Licensee shall also be entitled to the use of PH's likeness for advertising and promo-

tional purposes upon Licensor's approval first being obtained in each instance, which approval shall not be unreasonably withheld or delayed. Licensor shall make every reasonable effort, in light of PH's busy schedule, at the request of the Licensee, to arrange for PH's cooperation for publicity photographs, launch parties, personal appearances and radio and TV interviews (which shall be included in PH's obligations of seven (7) and four (4) appearances discussed above). Licensee shall reimburse Licensor for the reasonable costs involved in providing PH plus one other individual, selected by Licensor, plus her Mother and Father if they wish to attend, with first-class travel, lodging, food and other related expenses mutually agreed upon in advance of each appearance attended by PH at Licensee's request. If PH fails to appear for a scheduled Licensor approved event, Licensee will have the right to deduct up to $50,000 of its non-refundable out of pocket expenses incurred in connection with each specific event from the Sales Royalty. The failure to appear at a scheduled event could have a material adverse effect on the Licensee's ability to market the Articles.

The contract is signed by Frank A. Buttacavoli, COO, CFO, for Parlux, and, under the phrase "acknowledged and approved," Paris Hilton.

When, with all the irony implicit in asking something you know full well the other person is contractually forbidden from answering, I put the question to Timiraos, he smiled and said, "Although I of course wouldn't comment on Sarah Jessica's specific agreement; sometimes, with some celebrities and with most licensing agreements, there are minimum payouts; that is, regardless of how the brand does or even if you don't launch it, you guarantee the person a certain amount of money."

I have no idea what Parker's specific terms are. Whatever Hess and Walsh negotiated, it assuredly worked well for both of them, and for Hess's client.

Being an agent is one of the stranger jobs in the world. I once spent an LAX–JFK flight talking to an ex-William Morris agent. He was a really nice guy. He'd gotten out of the industry. He said, "Look, it comes down to the fact that you're living and dying by your clients, who can leave you at any second. For any reason. They have a bad meal, you get a phone call from their assistant. I couldn't do that, in the end."

We talked about Toby McGuire's brutal decampment from his ex-agent. Obviously if I ever asked SJP (which I never would, it's stupid), "What do you think of Peter?" she's going to say, He's great, he's wonderful. Why the hell would she say anything else to me? But the fact is that Hess was integral to making *Lovely* happen, and it wasn't in any way a foregone conclusion, and after talking to people about how tough it actually is to get a celebrity perfume launched, and the risk involved, and the negotiations and second guesses and all the other stuff he dealt with, I feel like I can smell his piece of the perfume too.

CHAPTER 10

BOUT THREE WEEKS before his deadline for finalizing *Nil*'s formula, I was eating with Ellena in a café in Grasse. His cell rang, and he had a brief conversation about the perfume museum he was helping establish in Grasse. It would contain great perfumes, old works of art now disappeared. And new things, perhaps like *Nil*. He hung up and, thinking about this, said, "One of the permanent worries I have is 'Am I still *en phase* with my era?' The tastes and the moments of the moment. The spirit of the time."

He took a bite, said, "I am of a certain age." (He was fifty-seven.) "At some point I'll have a vision *passéiste* of my time, not of now. *J'ai toujours la trouille.*" I'm always scared as hell about it. "Example," he says. "In my time I've been a head perfumer. So I managed people, and when they showed me things I'd say. 'This smells 1960s,' and when I smelled the older era I was scared because I could become that."

Perfume as a true art form has only existed since 1881 with *Fougère Royale* and its first use of a synthetic, but it is startling how fast the aesthetics change, the fashions blossom and die, the hot trends autoterminate in a concrete wall, surpassed by other trends. Ellena's fear is one every perfumer—every movie director, every composer—lives with. A perfume can come out, hit the List, be ubiquitous, and in the blink of an eye become not just unwearable but unthinkable. Go smell Yves Saint Laurent's *Kouros*. *Kouros* is a perfume that falls into the category animalic.

"Animalics" are animal smells. Castoreum, a perfumery material that comes from an abdominal gland of the beaver (*le castor* in French), smells of hot, worn leather, urine, smokey tar, and of anus—not (and this is very important) exactly feces but rather the smell of dirty, smelly, odorous, hot ripe flesh washed in fecal matter and sweat. It smells (in my opinion; perceptions vary) similar to labdanum, which comes from a bush and yet smells like the dirty testicles of a rutting beast. (That a bush can produce a substance that smells like an anus should not be surprising; as Ellena pointed out in Egypt, indoles, the smell of corpse, are loaded into jasmine, a flower that drips with the scent of rotting death.)

Civet, from an anal gland in the civet cat, is the fundamentalist's animalic, rawly, violently dirty, dark as pitch, strong as a hammer, and even more fecal. Ellena told me once that in his view civet contains an angle of the smell of blood, though personally I don't smell the blood. To me, the smell of blood = cold steel + cold dry cement.

But Tonquin musk is animalic in its most elevated form. It is a perfumery raw material that was extracted from a gland under the lower stomach and before the hind legs of the male of the species *Moschus moschiferus* L, the Tibetan musk deer. Not muscone, the molecule found at 2 percent inside this stinking cream; Tonquin musk is the real, natural, glandular product. It is one of the most astounding smells you will ever experience. It is, to put it most precisely, the rich, thick scent of the anus of a clean man combined with the smells of his warm skin, his armpits sometime around midday, the head of his ripely scented uncircumcised penis (a trace of ammonia), and the sweetish, nutty, acrid visceral smell of his breath. There's simply no other way to describe it.

(Some use *animalic* to describe ambergris, which is rancid whale vomit and a classic perfume ingredient and gives a marvelous, mesmerizing oceanic/proteinacious/creamy/salty scent, but they are deluded.)

The smell of clean anus turns out to be extremely helpful in perfume. In trace amounts it deepens and enriches floral scents, fleshes out green scents. Jacques Guerlain—this is a man who was creating perfumes as recently as the 1950s—famously said that all his perfumes contained, somewhere inside them, the smell of the underside of his mistress. He was referring to all three holes. Ellena uses animalics. When he is working his olfactory magic in front of you, he will dose it perfectly, so that you don't smell the animal per se but rather its effect on the whole. He'll do his trick with the isobutyl phenylacetate (the sweet/chamomile/vaguely chemical)+the ethyl vanillin (the rich gourmandy vanilla molecule), where you hold them under your nose and chocolate appears in the air. And that's when, with careful timing (he is a showman and enjoys making you gasp), he hands you a third *touche* of natural civet, and you will smell a mouthwatering unctuous chocolate ganache. The smell of shit is crucial to any high-quality chocolate scent.

But as a perfume? *Miss Dior* was created in 1947 and Caron's *Yatagan* in 1976. Both were huge hits, and both are animalics—*Miss Dior* smells like the armpit of a woman who has not bathed in a week, and *Yatagan* is the odor of a European man removing his underwear in August—and they are now categorically unwearable except by the French. Today *Kouros* will get you expelled from a restaurant. It was launched in 1981. It is brutally not *en phase*.

But Ellena will call into question even the most accepted conventions. In his opinion, the musks used in detergents fundamentally divide perfumery into philosophical points of view. "There are two great poles of perfumery," he says. "Latin and Anglo-Saxon. Seduction and hygiene. The Latin wants to seduce; he says, 'See how sexy I am, I'm coming to you.' The American says, 'See how clean I am, you can come to me.'" His reaction also results from his impatience for anything too easy. "Today if I want to please Americans, I make it thirty percent clean notes just automatically. I know they're going to say, 'Oh, we love it!'" He rolls his eyes, proposes (after a brief pause, and very dryly), "*cK One*" and then a moment later, "*Obsession*," and looks away into the middle distance.

(Incidentally, perfumers still do keep a few animalic synthetics in their palettes, though your chance of ever smelling one decreases every day with the rise of the Asian market, where the demand is for scents that smell of more or less nothing at all and where the police would be called at the first trace of civet. And yet the smells sit there on the perfumers' shelves in their little glass vials. In 1915, Sack took natural civet and from this reeking cauldron of hundreds of stinking molecules, he chemically tweezed out a single ketone molecule that is manufactured and marketed under the name civetone. It is a musk that has a hugely strong stink of body odor and anus. More powerful than this, however, was Walbaum's work in 1900 in Leipzig. He took civet, the cream from the anal gland of the civet cat, and from it excised another single molecule; it was given the rousingly poetic name skatol from the Greek word *skatos*, excrement. In fact, in 1877 Brieger had taken the trouble and done the chemistry necessary to isolate the molecule in human feces. Skatol is a molecule that incarnates the adjective "animalic.")

I asked Ellena: Do you have a good sense of the smells of various eras? "Oh," he said raising a cup to drink, then, "of course, each time has its smell. Sixties fragrances are structured in a sixties manner. Eighties in an eighties manner, and people will understand the codes you've put in. And this has nothing to do with the materials. It has only to do with the way in which *I use* the materials."

"Take men," he said. "The 1960s began a romanticism—consider the culture at the time—with fresh, big colognes, citrus, the epitome being *Eau Sauvage*, 1966. You can smell how the 1970s scents followed logically. The epitomes of that era, *Paco Rabanne Pour Homme* and *Azzaro Pour Homme*, which were heavy on rosemary, lavender, thyme, the smell of the Mediterranean, but done in such a way that made the seventies really *'parfums de camionneurs,'* truck driver scents, the guy who climbs down from some big cab. *Brut de Fabergé*. 'Men dare perfume themselves!' 'Virile, not elitist!' " Ellena says all this straightforwardly, with no irony but at the same time with an academic detachment that renders ironic.

"1980s: *Cool Water* and *Eternity*, both 1988, and they were exactly the same thing, though you're not supposed to say it. Clean, sterile, hygienic. That really shook up the perfumers because now we were using different codes and politics. You can never forget that perfume is completely political. This hygienic school was completely led by the Americans."

He changed the subject. "Hermès has let me talk to you, about whatever I like; I'm on my own. No one has said to me, 'You can't tell Chandler this or that.' I like Hermès for that."

He didn't, he said, referring to his new position, care about the envious and the jealous. Too bad for them. "What I find perhaps interesting is that by my deciding to go to Hermès, there is a small chance that maybe everyone will be able to do perfumery differently. I have an ego like everyone. But to say this in the most generous terms, I hope it will help all perfumers be more free."

⤳

As we were paying the bill, Ellena said, because the thought came to him, "I don't show them [Gautier and Dubrule] the formula because they wouldn't understand it. In the new museum, I'm going to ask them to show a perfume formula. The public won't understand, but at least they'll see there are real people behind perfumes. They say

THE PERFECT SCENT

cooking is an art and pastry is a science. Perfumery is a math, specifically an algebra. All these interactions." He moved his hands and fingers to indicate variables flying about, crashing into one another.

⌇⌇

The construction of a perfume bottle is a project far more complex than it appears. Years ago, Ellena had been directly involved in the creation of one of the more singular perfume bottles on the market. Its story illustrates the problem.

Around 1990, Ellena began talking with a cousin of his by marriage, an engineer named Thierry de Baschmakoff. Baschmakoff, descendant of a princely Russian family in exile that landed in France and born *grassois,* had like Ellena been raised more or less perforce in perfume. (Baschmakoff's brother-in-law is also a perfumer.) He is relaxed and friendly with an appealing demeanor, like a Californian who got waylaid accidentally in Europe.

Baschmakoff saw Ellena at family gatherings and so on, but he wasn't particularly interested in fragrance. What did interest him, in his smart-young-genial-guy way, were industrial design and technical problems, and some of the most interesting problems posed to industrial designers are those arising from the endlessly variegated creation of perfume bottles, where the stakes are high and the aesthetics intense. So he simply decided one day that he would leave his boring job—he was doing technical coordination of building projects on the Côte d'Azur—move to Paris, and start to design perfume bottles. (Most bottle designers sculpt their models in clay. He had never sculpted before and wasn't certain he could do things that small, but that didn't stop him.) He created an atelier, which he called Aesthete, asked Ellena for his contacts, and started looking for work.

There was none. Since the 1960s, four designers had entirely dominated the market. The two biggest were Pierre Dinand (who created the *flacons* for *Opium, Eternity,* and on and on) and Serge Mansau (bottles for Kenzo, Dior, Givenchy's *Organza*), the latter a fat megalomaniacal cigar-smoking blowhard whose pomposity and arrogance were as legendary as his habit of talking about himself in the third person: "Mansau makes art, yes, it is true, but we must ask ourselves: what *is* this 'art' of his?" Baschmakoff realized he'd been a bit precipitate. For five years he worked very hard and got almost no commissions at all, and finally he met some people from the Italian jeweler Bulgari, and there was at last a connection, and he did his first bottle, *Eau de Bulgari*, in 1992. He designed a single, strong, graceful sea-green column of sanded glass, tapering up to smooth, clear shoulders, and everyone looked at him differently after he'd done it. And there he was. He went on to create all the bottles for Bulgari as well as bottles for Cartier, Versace, Dior, Fred, Anna Sui, Yohji Yamamoto, Nino Cerruti, and Nina Ricci, but the *Eau de Bulgari* bottle was seen everywhere in the media and launched him.

Ellena did the Bulgari juice. Their collaboration worked beautifully. Baschmakoff got an idea: He proposed to Ellena that the two of them create a perfume company.

It would be a somewhat different company in that Ellena would do all the fragrances, Baschmakoff would create all the bottles, their juices would be extremely expensive, and they would not be owned by Lauder or L'Oréal. They called it The Different Company, and it was Ellena's position of in-house perfumer that he passed to his daughter, Céline, when he left it for Hermès.

Creating the house was high risk. Due to Ellena's reputation, they got it funded, and it opened its doors in 2000, and Ellena created four juices: *Osmanthus, Rose Poivrée,* which is the perfume Satan's wife would wear in hell (it is an exquisite scent, a combination of rose and smokey fire), *Bois d'Iris,* and *Divine Bergamot,* which recalls Jacques Guerlain's quote about putting into his perfumes the smell of the underside of his mistress: *Divine Bergamot* smells like the mistress sitting on an exquisite bitter orange. The Different Company's scents are sold in places like Estnation, the extraordinarily expensive, fashion-forward department store in Tokyo.

Baschmakoff wound up deciding to do a single bottle for all its perfumes. It is extraordinarily expensive.

The bottle is both a technical and aesthetic obsession for the perfume industry, in particular its marketers. (This is viewed a bit cynically by the perfumers, one of whom commented to me acidly, "Oh, of course, the *bottles* are visual objects, so the creative executives feel competent to control them. The invisible smell? They're lost, terrified. The bottle reassures.") The industry is conscious of the degree to which the bottle's presentation of its juices determines the fate of those juices. On the other hand, it is conscious of how much those bottles cost. Shiseido's mythic *Nombre Noir* was presented in such sumptuously luxurious packaging that though known as a spectacular perfume, it became equally well known for the amount of money it supposedly lost due to its wrapping.

The financial constraints imposed by the luxury brands on everything from the juice to the bottle are heartily lamented by the artists. The licensors "rarely ask you for beautiful products anymore," said one bottle designer. "You're always using the same materials. Think of the things we could do with a little more money!" Contrarily, in Baschmakoff's view the bottle designers sometimes have more liberty than the perfumers to express the brand, though if you press him he'll admit that exceptional things are rare because there's always a limit on the price. "Our work," he says grimly, "is very *parametré*" (parametered, constrained) "by the fashion houses."

The negotiations over cost begin the instant the house shares its vision. They want a bottle like this, the wildest fantasy, insane beauty, et cetera. The designer replies that it is, in fact, not technically possible to create a bottle like this, but he could do that, and it will cost X. The house recoils, demands price Y; the designer concedes on a few points; the house is still shocked, then lays down price Z that its bottle must not exceed; the designer replies that he cannot do it for that price. They wind up compromising on a cap in metal, but they'll drop the round edges because the consumer won't see them. Truly round glass edges are expensive. Truly ninety-degree-angled glass edges are more expensive.

The Different Company has proven the exception that confirms the rule. Its artistic freedom came, rather ingeniously, from the simplest of strategies: The two partners decided that Ellena would quite simply create juices too expensive to be attempted by anyone else. The price would automatically be immense, which meant protection from competition at this level. It also meant creative liberation since, say, Calvin Klein could never dare to follow into this stratospheric territory. In *Bois d'Iris*, Ellena specified an immense amount of Florentine Iris Root Butter, which costs, depending on the quality, the source, and the day, from $25,000 to $80,000 per kilo. Though at the same time, his formulae were as always relatively streamlined *"parce que,"* Baschmakoff explained, *"Jean-Claude n'aime pas les parfums compliqués."* Because Jean-Claude doesn't like complicated perfumes. Baschmakoff applied to his bottles precisely the same principle. He would create his flacons using exactly the quality of materials he wanted and generate the price, whatever it was, forward from there. (As opposed to the habitual inverse.)

In the event, the Different Company bottle is a work of art so gorgeous, so substantive, that Ellena's juices must do a bit of throat clearing to get the buyer's attention. The real cost of the 90 ml bottle is €2.5; of the 250 ml, €6. Some of this is simply due to the fact that once all the variables are in place, the price of each bottle will fluctuate wildly depending on the number in your lot. If you are the Different Company, you order 5,000 bottles, giving you that per-unit price. If you are Bulgari, you order 200,000 bottles for *Bulgari Blu* and each one comes in at between €0.40 and €0.60. If you are Estée Lauder, you order a million bottles per year, and the price is risible, maybe €0.25.

At Hermès and every other house, when you start creating your bottle, you immediately begin making choices. Every choice is both aesthetic and economic. Every bottle first has to be modeled. That's usually around €150,000, and there are no economies of scale; if you've got two sizes of perfume—a 50 ml and a 100 ml—it's €300,000. You can do a simple artwork on the bottle for €0.05 or a complicated decor that adds a full €1. You can use *pistolétage*, an enameled paint that can cost a little or a lot, a special *pistolétage* called *nacré*, or *strass*, plastic jewels that are—intentionally—the height of kitsch. (Unsurprisingly, several Versace bottles use *strass*.) *Strass* can add an appreciable amount to the cost because a robot must glue each jewel onto the neck.

Then there is the cap. Plastic (inexpensive), solid metal (expensive), or a treatment with a mechanical process called *décoltage* (very expensive). For economy, do a metal cap with plastic injection and *un lest*, a weight buried inside to fake the sensation of heft. Your tube and pump, which connect to the cap, usually cost around €0.15 for 200,000 units minimum, and you can cloak the tube in TPX (polymethylpentene plastic), which adds around €0.10 per tube. One executive told me, "Just the fucking pump for Gaultier's *Fragile* cost €200,000 to develop because they wanted a pump you can use from the bottom. And that kind of special-order stuff isn't all that rare."

(And then there is the packaging the bottle comes in. "The design of the packaging is an absolutely huge investment," says a French executive. "Everything has to be modeled. For the bottle you have to invest $150,000, and if you've got a 50 ml and a 100 ml, then it's $150,000 by two. There's stock packaging. You'd be surprised how much stock packaging costs. I'd rather not give examples. Still, it's a good business decision, the box is no big deal, but people are very sniffy about it.")

The meeting in Hélène Dubrule's office on the flacon (the bottle for *Nil*) took place on June 9 at 11:30 A.M. Dubrule faced the problem of what to do with her bottle for *Un Jardin sur le Nil*.

Hermès had endured a long, slightly tortuous experience with *flacons*. Dubrule and Gautier had talked about it. This is what they'd had: The bottle for *Calèche* (1961) had been created by Annie

Beaumel. It was a simple, elegant vertical sheath of glass with rounded shoulders, quintessentially good taste, and had been used for almost every perfume, the sheath astutely quoted: For *Hiris*, the sheath had been done in blue; for *Orange Verte*, it was tinted green and scaled down. Hermès had departed from this design only four times: *24 Faubourg, Rouge, Rocabar* (which was supposed to be shaped like a horse blanket), and *Eau des Merveilles*.

Gautier and Dubrule had decided that the Annie Beaumel— designed bottle really was the best physical representation of Her-mès scents, and they'd chosen it for the *Jardins* collection. This was wise in terms of the economics of the thing (no development costs) and the marketing, or, as the specific phrase went, "brand coherence." But if they would now adapt it to *Nil*, the trick was in capturing the look of the lotus motif in this piece of glass. This was what the meeting was about.

Dubrule ran it, a quietly taut session at the round table in her office. Anne-Lise Clément, the perfume group manager, was on one side of Dubrule, and on the other Guirec Boidin, her assistant head of product, young, blond, and always looking freshly show-ered. Across the table from her were Philippe Bleuet, the director of purchasing and development, and Virginie Lamejer, the head of packaging.

Bleuet and Lamejer were staring critically at the flacons Clé-ment was briefing them on. They were debating four proposals from the French glassmaker Saint Gobain. They had done three different mock-ups using the 100 ml bottle and a fourth mock-up with the 50 ml. Saint Gobain had done very subtle, almost trans-parent *pistolétage* on each of them. The trick was to make the colors blend correctly, not too much green, not too little blue. One of the 100 ml bottles had an aqua blue bottom, another green, the third yellow, and the 50 ml had yet a fourth distribution, intermediate aqua-esque coloring, each one carefully worked out.

They needed the bottles by August. So they were (a bit tensely) looking at June 15—six days from now—to launch the final orders on them, "which was late," noted Clément; Dubrule pursed her lips and asked everyone their vacation dates during the upcoming August shut-down of France. She wrote down each person's absences and glared at the dates for a moment as if they themselves were impeding the process.

Back to the bottles. Saint Gobain was using a glass *vernis* (the technicians would regulate the paint jets to produce precisely the effect they'd choose) dried by UV light and then baked.

Clément: "The fifty-milliliter coloring is the freshest."

Dubrule (eyeing the bottles): "There's less of a green background in the fifty."

Clément: "We found that when we forced it, the color flipped and we got a gray effect. See here"—she picked it up—"you've got this grayness? On the bottom?"

Dubrule pursed her lips here. "I don't see it," she said squinting, then dismissing it for the moment, "We have to see what it looks like next to the box." She reminded them that June 21 was the deadline for the final box mock-up.

Bleuet was flipping crisply through pages: "On scheduling, we're looking at the fifteenth of June to launch the orders so as to have the bottles by August." After discussion, they decided that, in fact, the 50 ml bottle really was perfect. (Clément and Bleuet looked subtly relieved.) In production they'd copy it to the 100 ml but "*forcer un peu la couleur,*" add more intensity.

Lamejer consulted her notes. She had long brown hair and a practical manner, and you could see her calculating the variables in her head. "The doc for the text on the bottles, 13 August. I can't do it faster than that." The text would be silk printing—*sérigraphie*—the silk a polymer, cooked at 600 degrees Centigrade. She pointed at the flacons. "The word *Paris*"—which would go below the words *Un Jardin sur le Nil*—"is always printed nineteen millimeters up from the bottom of the bottle," she said to them. "*C'est la règle.*" That was the rule with every Hermès perfume. This was so that when the Hermès store manager stocked a bottle of *Nil* next to a bottle of, say, *Eau des Merveilles*, the logos would always be at the same height.

They turned their attention to the *étui*, the box, as its colors had to match those of the flacon. They discussed what color yellow de Mareuil used in her leaves. "Jonquil!" said Lamejer. They were actually a rather khaki green. Dubrule touched one: "A bit too red here . . ." She pointed at a tiny dot on an étui leaf, pointed at another tiny dot, pointed at a tiny dot of black.

Clément shook her head. "The problem is that the blue turned green when we added more black here."

Bleuet: "We have to rework it in the real tint."

Boidin looked at the 50 ml bottle. "It's going to be slightly more difficult because it'll be two scans."

Dubrule: "Anne-Lise, you verified this model in natural daylight?"

"We're going to verify it," said Clément. Moving on, she got out some perfumed incense sticks, the first samples sent by the supplier. These actually weren't perfumed with *Nil*; the maker had sent them simply so they could look at the sticks' length. (Dubrule absently smelled one, frowned, murmured, "Wait, *what* is this scent . . . ?" She put it down, and Boidin picked it up with an interested, tentative expression. Smelled it. "Oh!" he said, frowning intently, "*what* is this scent . . . ?" They stared at each other.)

"He can do the right length of sticks for the *coffret*," said Clément, referring to the incense manufacturer. "They'll be wrapped in tissue paper so they don't move around. Notice they won't be the full length of the box because we need to leave room for the tissue paper that will wrap them." (They were extremely fragile.) "Or transparent paper or cellophane—we don't know yet."

Dubrule changed gears with an intake of breath and a look: "OK! So now we do our big scheduling!"

The 2005 theme of the year—rivers—would be announced in

January, but they'd reveal *Nil* to the press in November 2004 so that the press would be playing a guessing game re the theme.

Clément: "The cap—weren't we talking about a metal cap?"

Lamejer (bluntly): "Metal will change the size of the box."

Dubrule: "Is it a problem for you to have two caps?"

Lamejer thought about it an instant, clicked her tongue no.

And the *sophistiqués,* what they called the minisprays that would be filled with *Nil* as *échantillons,* samples. This was a possible headache caused by chemistry: The question was what would happen when Ellena's molecules came in contact with the bottle plus the tube and the cap that Dubrule was designing—or with the alcohol, or the body lotion base it would be mixed into? *Sophistiqués* and perfume bottles must be able to go from Iceland to Saudi Arabia, so Ellena's juice would be heated to fifty degrees Celsius and frozen to below zero, baked and shaken and have an electric current sent through it, then analyzed for any alteration. Aside from already being late, the five at the table were acutely aware of what had happened with Ellena's first *Jardin* juice; with *Méditerranée* they'd had to throw out all the *sophistiqués* in plastic—standard industry *sophistiqués* used for any number of fragrances—and hastily replace them with glass because his formula had interacted with the plastic. No one knew why, it was impossible to predict; it happened rarely, but it happened, and what was strange was that any two molecules together rarely interacted; each formula was its own question mark. And the immediate problem was that Ellena's formula for *Nil* was not yet finalized and thus couldn't be tested. The question was: Would the plastic change the smell?

And in the factory they'd put in the preservatives, which virtually all perfumes have, antioxidants (oxygen atoms damage perfumery materials) and radical scavengers like butyl hydroxy toluene (BHT) and benzophenones. (BHT has been used for many years with good success and is added to many essential oils. Benzophenone is a ketone with a faint rosy/metallic scent that can be used as a perfumery material for its smell. It is also edible and so can be used to make flavors. When used as an antioxidant it's dosed at such low levels that it has no effect on fragrance.) And the more advanced anti-UV sun-filter molecules they were starting to use (the industry calls them UV absorbers), basically sunblock for perfume, like a mix of Parsol MCX (2-ethylhexyl 4-methoxy cinnamate) and Parsol 1789 (4-*tert*-butyl-4'-methoxy dibenzoylmethane) or you could try Tinogards like Tinogard TT (3,5-bis[*tert*-butyl]-4-hydroxy benzenepropanoic acid). They prolonged life from a few months to a year, but it was tricky because each fragrance had to be fitted with the sunblock that worked for it. The ideal was to have an opaque bottle, like the one Tom Ford used for *Black Orchid*, and the truly ideal was to store the perfume in the fridge, but this stuff helped; take one scent with, one without, let both stand in normal sunlight, and, a perfumer would tell you, you can smell the difference after a single month. (There were the inevitable ideological objections to the preservative molecules, but, noted a perfume chemist, we swallow $C_{17}H_{18}F_3NO \cdot HCl$—aka Prozac—and 2-(4-isobutylphenyl)propionic acid—ibuprofen—via our digestive systems, which send these molecules to our brains. We have botulinum toxin type A—Botox—injected into our faces. But we don't want BHT sitting on top of our skin? "It is," said the chemist, "irrational.")

Moreover, the decomposition products caused by UV radiation can cause allergies or, worse, be toxic, which means that contrary to the claims of the all-natural movement, preservatives actually preserve health.

The chemists would use one of a number of alcohols—40B, 38C, et cetera—the only difference being the denaturing agent, plus half a percent of water. "Denaturing" alcohol, which is ethanol, just means adding something to it that makes it undrinkable—put in phenolphthalein or isopropanol, which give a terrible taste. Otherwise alcoholics would drink perfume. It's actually an important financial consideration: There's no import duty on denatured alcohol; if it weren't denatured, a bottle of *Un Jardin sur le Nil* would have the added cost of the taxes on Stoli.

The *sophistiqués* were a problem because they had to go out first. The deadline for the *dernière validation*, the definitive juice, was July 15. Which was late, said Clément with a sigh.

Bleuet: "We can gain some time if we don't do the preseries for the press of the *lait de corps*," the body lotion. It needed to arrive four months before being put on sale; this was another problem with chemistry: The molecules in the *concentrée*, the pure ingredients in Ellena's formula, needed at least six weeks of maceration in alcohol to mature its full scent. "We'll just do the *parfum*."

Dubrule: "Excellent idea. The 10 July, validation of the *jus*, the documents for the *étui*, 13 August at the latest. And we have to pressure them for the list of ingredients for the incense. It's 9 June today."

Dubrule was unconsciously waving one of the incense sticks in the air for emphasis, a conductor with a tiny baton before an invisible orchestra. She suddenly froze. She held the stick under her nose. "Ah! I know what this is: *fleur d'oranger!*" Orange blossom.

She gave it to Boidin, and he smelled it and smiled and nodded vigorously: Of course, it was a wonderfully fragrant orange blossom scent. They were relieved to have figured it out. Nothing makes you crazier than to have before you a scent that you know perfectly well and not be able to identify it.

⁓

Ellena was in Grasse, working and thinking and evaluating some new molecular machines he'd created.

With an eye to future Hermès fragrances, always searching for new scents to create new olfactory sculptures elegant and (he hoped) marvelously strange, he was working with Frédérique Rémy and LMR to create literally new materials—tailor-made molecular distillations of naturals, designed to his specifications. Rémy had just sent over a dozen of these for his perusal, and he opened them at his desk. Hermès was one of the few houses that could shoulder the expense, but the result was worth it. Under his direction, Rémy's technicians were teasing out very specific molecules found in natural rose, violet leaf, and summer straw, only the molecules Ellena wanted (they dropped the others on the cutting-room floor), thus making an Hermès rose, an Hermès violet leaf, straw, something new under the sun—not absolutes; not essences; not naturals; not synthetics.

This is what one molecular distillation of Turkish rose absolute looks like, and it is thus only eighty-some molecules; the list of the total, unedited molecules contained in Turkish rose would give you between eight hundred and a thousand different molecules, depending on the species, and go on for pages. This analysis was done by GC/MS. This rose—not synthetic, but not natural either, rather a hybrid in-between—is a carefully engineered scent, a rose greatly (very greatly) simplified into a flower that doesn't exist in nature but is, for being engineered by us, even more naturally spectacular. Amounts are in percent of formula. Notice that none of the ingredient names is capitalized (except the acronyms) since only commercial trade names of synthetic molecules are capitalized (Exaltolide) while natural molecules are lowercase (citronellol), and these are all molecules built in flowers by nature.

citronellol	45.07
geraniol	17.91
nerol	9.60
nonadecane	3.21
eugenol	2.98
PEA	2.67
linalol	2.14
heneicosane	1.33
alpha-pinene	1.08

geranial	0.99
geranyl acetate	0.91
citronellyl acetate	0.76
heptadecane	0.73
nonadec-1-ene	0.72
caryophyllene	0.63
rose oxide I	0.58
alpha-terpineol	0.57
isoamyl alcohol+ 2–methyl butanol	0.56
eugenol	0.56
terpinen-4-ol	0.52
phenylethyl-acetate	0.52
germacrene D	0.43
hexanol	0.43
neral	0.33
alpha-bulnesene + pentadecane	0.30
eicosane	0.29
tricosane	0.29
alpha-guaiene	0.28
farnesol-trans-trans	0.26
alpha-humulene	0.25
rose oxide II	0.25
beta-pinene	0.24
myrcene	0.18
neryl acetate	0.17
heptanal	0.16
nerolidol	0.13
isovaleraldehyde	0.13
nonanal + hexanal diethyl acetal	0.12
heptanal diethyl acetal	0.10
farnesol	0.10
gamma-terpinene	0.08
limonene	0.08

2-(2-methylamyl)-4-methyl-2-3-dihydropyran	0.07
methyl-geranate	0.07
pentacosane	0.06
cis-3-hexenol	0.06
p-cymene	0.06
amyl alcohol	0.06
ethyl decanoate	0.06
heptadec-1-ene	0.05
linalol oxide I	0.05
citronellal	0.05
hexanal	0.05
beta-bourbonene	0.05
docosane	0.05
beta-elemene	0.04
octanal	0.04
octadecane	0.04
phenylethyl-isovalerate	0.04
phenylethyl-tiglate	0.04
methyl heptenone	0.04
linalol oxide II	0.04
phenylethyl-benzoate	0.03
delta-cadinene	0.03
ethyl hexadecanoate	0.03
sabinene + benzyl methyl ether	0.03
hexadecane	0.03
eicos-1-ene	0.03
heneicos-1-ene	0.02
octanal diethyl acetal	0.02
ethyl dodecanoate	0.02
1,8-cineole	0.02
trans-ocimene	0.02
phenylethyl hexanoate	0.02
dimethyl styrene	0.01

benzaldehyde	0.01
cis-ocimene	0.01
methyl salicylate	0.01
ethyl benzoate	0.01
phenylethyl-isobutyrate	0.01
Total	100%

Ellena was creating these, peering into flowers and leaves and roots and choosing exactly what he wanted, and Rémy's technical wizards were then going in with their scalpels and slicing out what he didn't. So he was designing new olfactory tools for himself, natural smells that had never before existed in nature, to build future perfumes.

Gautier and Dubrule had decided that the European press launch for *Nil* would be November, early December in New York, and the product needed to be in the U.S. stores in February. Summer was moving much too quickly. They'd set the deadline for final acceptance of the juice for mid-June, now long past, and so—new deadline—Gautier stated she would need to make a final decision on the formula by the beginning of September "at the very latest."

In July, Ellena went back to Paris for the next submissions.

From AG2, AJ1, AJ2, AJ3, he had come up with AJ and shown it to Dubrule and Gautier. They'd been putting themselves through the usual anguish over it: Was AJ too cologne, too citrus, too *eau fraîche*, too bergamot? Was it not sufficiently "perfumey"? Due to years of acculturation, primarily from *Chanel No. 5*, people had come to think of a "perfume" as a juice that had aldehydes, the powdery soapy smell. Ellena and Gautier had to think about the preconceptions of the market. AJ was a completely different beast from the scarlet Parisian boudoir vanillic plush of *Shalimar*, the French-fundamentalist animalic of *Miss Dior*, or the light-gold-evening-gown paradigm of Jean

Patou's *1000*, that wonderful rich, hard-as-diamonds elegance that is like a spotlight at a Cannes movie premiere. Nor was it a contemporary electrical, like the Hilfigers (some of which are excellent, Calice Becker's *Tommy Girl* being a masterpiece of the genre).

It had no tired clichés like dihydromyrcenol, the smell of bland, freshly showered jock, but on the other hand, the clichés were easily accessible and thus easy to sell. It had the beauty and presence of Givenchy's glorious *Amarige*—which is one of the greatest formal scents and manages a zipped-up elegance without stuffiness. But it didn't have *Amarige*'s wonderful artifice, the sequins, the sense of being constructed, no olfactory luxury logo in gold lamé letters sewn into the silk lining. Which brought them back, finally, to the same question, and now it came down to this: They smelled it again and, a bit grimly, acknowledged it was "not perfumey enough." AJ was not their *Un Jardin sur le Nil*. Ellena needed to do more work.

OK, so where were they right. They did want to keep AJ's delightful (as Dubrule said) *fraîcheur vivifiant*, its enlivening freshness, and at the same time give it (Gautier on this point) *plus d'épaisseur*, heft, body.

So Ellena flew back to Grasse and got to work. He modified ingredients, put things in, took others out, and got on a plane back to Paris and presented them with the next modification: AS.

AS was much more flowery/fruity, a bit sweeter, a bit more ripe mango. Gautier didn't really like it, finding it too easy. A bit too "sixteen-year-old girl," she said. Dubrule, on the other hand, found it quite good, then tried it on her skin and pronounced it in fact too sweet. It lacked a certain elegance in her opinion.

Ellena got on a plane, flew back to Grasse, and thought and drummed his fingers on his desk and balanced milliliters of molecules against other milliliters of molecules. He began creating A*.

A* began a gentle swing toward the woods. He was editing and rewriting the formulae, draft by draft, sending each draft to Monique, his lab tech, like sending in a photographic negative, and then looking carefully at the colors and details of each developed photo she sent back to him. He lowered the natural citrus and excised a third of the bigarade, the bitter orange, then he welded on three Monique Rémy materials: an incense at €150 per kilo (a material that smells of summer camp: fresh mountain + pine tree), a bigarade essence (usually €100 per kilo, but Frédérique gave it to him at €80), and an *absolu de miel* (honey absolute) at the rather significant price of €1,629 per kilo. If successful, *Nil* would generate annual sales for Monique Rémy of 500 kilos of bigarade, 50 kilos of incense, and 50 kilos of honey; Rémy would use 10 tons of bitter oranges to obtain the 500 kilos of bigarade, and she would import them from the Ivory Coast, unless there was a problem, in which case she would go, reluctantly, to Italy. The Ivory Coast quality is excellent, and the production cost is of course much lower. Italian oranges are twice the price for, in Ellena's estimation, an only slightly better scent.

He smelled a bit and thought about it, went home and had dinner and didn't think about it. Came back to the lab and added an expensive synthetic that hadn't been in AJ called ambroxan, made by the German company Henkel, a lovely wood/musk that cost €500 per kilo, as well as muscone (another costly synthetic; he wanted it in this case mostly for the *persistance*, for helping the perfume last on skin, rather than for its scent character). He added a very inexpensive IFF synthetic called Iso E Super at €15 per kilo, a beautiful woody cedar note, very diffusive, terrific for the *sillage*. (Lauder's *Beautiful* is powered by Iso E Super.) And he used the new Hedione. Hedione was a 1970s synthetic, one of Firmenich's fabulous captives, these new patented molecules that took the world by storm, but Hedione Supérieure at €150 per kilo (ten times the price of the original) was the company's new captive. It was more a white flowers scent, whereas the original molecule smelled flowery/citrus. The Supérieure version had more *épaisseur* and would help, to use Ellena's phrase, *donner du gras* to A*. Give fat.

In *First*, for Van Cleef & Arpels, he had used 160 ingredients. There were exactly 30 in A*. He smelled it. It was green mango, the very instant before being picked. But not the fruit, exactly. He'd wrapped it in both a fresh water and a warm, clean desert air, and you smelled Ellena thinking about the mango, on the tree, above the sandy dry road. It was an idea he had created of the fruit, the scent abstracted, fed through his mind, which registered all the loveliness, the limpid quality of the one moment in one season when, if you were in the right place at the right time, the scent existed, unripe, just before disappearing. Beautiful because ephemeral.

⁓

On August 16, he went back up to Paris, and, in the offices at Pantin, he presented A* to Gautier and Dubrule.

Days passed. He waited. It was the usual—he wondered when he was going to hear, wondered why Gautier didn't simply take it because he knew it was terrific, knew why she didn't take it (because it didn't work at all, because he'd forgotten to do this material, to do that one, to balance better that other thing; "Those are the truly unpleasant moments," he said). At other moments he actively didn't think about it, or tried not to. Gautier, he knew, was someone who, even as the deadlines raced closer, believed herself obligated to ask questions as long as she possibly could—Can I make it better? Can I go further?—and the decision to lock in the formula, on a deadline hopelessly receded, was clearly going to be made at the very last second. Bernard Bourgeois at the Normandy factory had to ensure correct maceration, which meant he needed the formula. Gautier would wait until the factory started screaming.

And when he wasn't actively not thinking about it, Ellena was tweaking the perfume. He'd put in things, take things out, ask himself constantly: Does it diffuse enough? Is the *persistance* sufficient? Mostly technical problems now. The aesthetics he'd already worried into the ground.

On September 7, 2004, at the last possible second, Gautier called him and took it. The scent was registered in the laboratory. A* was *Un Jardin sur le Nil*.

~⊚~

It is a perfume that smells like early evening on an island where it is always summer. It is the smell sunlight makes coming out of a blue sky, the air scented with the tang made as the light warms the smooth unblemished peel of the greenest mangoes hanging from the branches of the young trees, just out of reach.

CHAPTER 11

B Y THE BEGINNING of February, Walsh and Timiraos face a decision: What version of *Lovely* will they put in the new product? Will they change it, inflect it with one of the scents the IFF team has shown them, or will they use Le Guernec and Gavarry's classic? After much discussion, they decide on the same formula. It will create continuity, keep the product pure.

They discuss it with Parker and lay out their logic. She agrees.

Their last meeting takes place on February 21, 2006. Parker, Walsh, and the rest of the team gather in TAGland, assembling around one of the big white lab tables. The conversations stop. There's an awkward moment.

"Our final meeting," says Timiraos after a moment.

"I can't believe it," says Parker. There's a sense of achievement and a wistfulness of the end.

Timiraos clears his throat. Right, first order of business: What name to give the product. He, Walsh, and Leslie Oglesby, the global marketing director, present the options to Parker. A significant amount of work has gone into this, first the creative—generating names—and then the slogging through legally trademarked names to find out what's been taken, what's protected, and what they can actually use. The Coty team is also a bit nervous (Parker is perfectly aware of this) because they have specific opinions, but she has ultimate control.

"You ready?" says Timiraos. Parker nods, eager to hear. "OK," he says. "The names *Liquid Silk, Satin Skin,* and *Skin Silk* were unavailable," he says, just to get them out of the way. She nods: OK, got it. "So here they are. *Silk Slip,*" he says, slowly and carefully, "*Satin Slip, Liquid Satin.* These are the ones that have passed legal." He hesitates, then can't help adding with a little regret, "We actually were loving the idea of *Liquid Silk* because fabric was so important for *Lovely.*"

Oglesby has to agree. "And the product almost feels like silk on your skin."

Timiraos: "But we love the top three. *Silk Slip,* which is a play on words, *Satin Slip,* and *Liquid Satin.* And"—he takes a breath—"we think any of those would work very well."

They wait for a moment, and the room focuses on Parker.

Timiraos begins, "Catherine likes *Liquid Satin*—"

Parker: "That's my first choice."

Timiraos: "Great!" He and Oglesby smile at each other, relieved, then both grin at Parker, who sends up their relief by saying very loudly, "'Oh, thank *god* she chose the right one!'"

Timiraos: "Some people think satin is a cheaper version of silk, but I don't at all—"

Parker, simply: "Not at all."

Timiraos: "Great!"

Now the box. Chad Lavigne has been working on mock-ups, and Parker moves down the broad surface of the table to the far end where Lavigne stands at laid-back, cheerful attention, all six-foot-three, tousle-haired boyish elegance of him, beside the neatly lined-up results of his work. She loves the four boxes, turns each around, picks them up, and talks about their details. (Walsh and Timiraos have, over months, spent hours with Lavigne, working through the iterations of each design, and what Parker is seeing is the final, winnowed result.)

Parker holds the first box and exclaims, "I love it, Chad, it's gorgeous." After a few minutes she settles on one in particular. "I LOVE this idea! I think it's thrilling." The box is a translucent acetate. She notes a seam (Lavigne leans in to frown at it up close, traces his finger along it next to her finger: "We'll get rid of it," he murmurs), points out a problem with the folds. Lavigne has mocked it up with "Sarah Jessica Parker *Lovely*" and then the name, carefully printed, "Liquid Satin"—he'd guessed right—underneath that. (He'll further refine it. In June he, Walsh, and Timiraos will change their minds and in fact decide against the translucent clear, and Lavigne will redo the design, working it into an opaque with a filigree "because we wanted it to pop more in the showcase," Timiraos will explain. "The clear box, it turned out, got a little lost, and it didn't look expensive enough.")

Timiraos puts a hand on my shoulder and leans in low. "I think she approved everything . . ." He's doing his job managing the journalist.

Well, I say, I'm writing it!

"Fine!" he says, and quickly announces to the room, "then it's official."

And that's it.

SJP leaves a little while later. Everyone sees her off as she crosses the West Fifty-seventh Street sidewalk toward the big black town car waiting for her. The driver opens the door and she gets in quickly, he runs around and gets in, and we can see her say something to him, then sit back. Her hand waves, and she's gone.

CHAPTER 12

THE MORNING OF September 24. Ellena's formula for *Nil* was in the process of manufacture, but this meant that now they had an entirely new deadline, the press launch of the perfume on November 14. The perfume was done, but they now had to deal with the logistics surrounding it, plan and deploy its entry into the market.

Dubrule went to the coffee room and got a tiny plastic cup of coffee from the vending machine. The coffee room was next to the smoking room, French workplace institutions that, by around 10:00 A.M., faithfully re-create throughout the country thousands of filthy little Dachau gas chambers. Leaving with her coffee, Dubrule ran into Gilles Coupat, the head of Hermès in France and Belgium. "Where are we?" she asked him immediately.

Coupat was negotiating with the new Galeries Lafayette, the primordial Paris department store just above the Opéra, which was in the midst of creating what it calls the largest perfumery in the world. He reported to her that Galeries wanted to put Hermès's *stand* (the industry term in both French and English) next to Mugler and Gaultier, which was fine, but also next to Serge Lutens, which was in a sense even finer—Lutens's stuff is niche luxury. The problem was that all four would find themselves in slightly out-of-the-way territory. Coupat wasn't quite happy. Dubrule pursed her lips. "I'm quite happy for them to present us as *haut de gamme*, top of the line, but there has to be traffic." Coupat replied, "We're working on it."

He changed the subject. The word was going around the office that there was some sort of delay with the juice, apparently. A problem at the factory? Dubrule replied that she had had a few brief conversations with people, but no one had spoken either with Ellena or Bernard Bourgeois at the Hermès factory in Normandy. Maybe the rumor was incorrect.

In any case, added Dubrule, she was dealing with a more fundamental problem. They had been aiming at mid-June to lock in the formula, they'd wound up doing it mid-September, and they had (she said this to Coupat with a grimace) been smelling it, and A* still wasn't diffusing exactly as they wanted. Also there was a bit of acidity, of *aigreur,* in the top of the scent, and in the bottom it still needed more *tenue,* hold. She and Gautier had had a slightly tense telephone conversation about it with Ellena, who'd said it was a matter of *en finialant,* fine-tuning, the thing. At least, Dubrule told Coupat with a sigh, they'd finished the video of the trip to Aswan, the book was done and printed, and after laboring over two proposed directions for the *dossier de presse,* the materials they'd be giving the media, they'd chosen one and were quite advanced on it. If the juice would just behave now.

Coupat went left to his office, and Dubrule headed right back to hers with the tiny plastic cup of coffee. In the corridor she ran into Anne-Lise Clément, who had some news. The communications giant Publicis et Nous was creating the launch's visuals, the images that would appear in Hermès's spring 2005 ad campaign—taking the photos, doing the layouts; Dubrule had fully briefed its staff on *Nil*, told them the whole story. The problem was that the visuals weren't ready. Dubrule absorbed the information, making a very terse roll of the eyes. "Well," she said. She visibly added it to her mental list.

She and Clément went into her office and directly into a meeting on the PLVs (*publicités sur les lieux de ventes*), the counter displays, window displays, and podium materials. Virginie Lamejer, Hermès's head of packaging, was there, and a few other Hermès people. They were meeting with Franck-Alexandre Basset, the very tall, energetic young director of creation for Carré Basset Associés, the consulting group on the project, who was sporting a chic haven't-shaved-in-two-days look and presenting several visual options for the PLVs, which were plexiglass displays that would showcase the perfume.

PLVs are a necessary evil, created to draw the shopper's fickle attention in other people's stores like Galeries Lafayette, where Hermès had little control over its products and was one more tiny plot of land (its stand), its perfumes competing in a vast sea of glass and chrome with the latest Guccis, Fendis, and Revlons. They're also used in the consumption halls of airport duty-free zones, where everything from whiskey to watches was shouting. "It's a sensory game," said Basset, playing with plexiglass display. (Hermès would never allow these things in its own stores. Hermès stores use discreet, simple decoration, and of course the products are 100 percent Hermès.)

Basset presented them with several visual options, clear plexiglass tubes and a triple-layer plexi shelf with multiple layers of paper lotus leaves; the shelf would hold the bottle in front of a photo of the Nile—he had used one with the details ghosted, another in sharper relief. They talked about which would produce more enchantment. He tried two leaves between layers of plexiglass. Dubrule warned "Two is not a good number in general." Basset:

"Ah, but it's two on the *first level.*" He played with it. She got up and started shifting leaves here and there, standing back and looking. "The idea," said Basset, "is to have the bottle floating above the lotus leaves. Where does the Hermès ribbon go, this way? No, put the bottle here."

He wanted the plexiglass to overshoot the base by a half inch, but that would make the thing more fragile, and Clément reminded them that store environments can be rather violent. "It has to create enchantment," said Basset. He turned the image sideways. "Perhaps this is more enchanting." (People in the luxury world, with no apparent irony, take quite seriously the idea that one can create "enchantment" with pieces of plexiglass and some glue.)

"There's still a problem with the thickness of the plexi plaques," said Dubrule.

Basset: "We can eliminate one." He talked about glue, on the plaques' surfaces, on the side that says "Hermès," and Dubrule asked if they could put metal tubes through the triple-layer lotus-imbedded shelf. Basset put the Hermès ribbon on the leading edge, and no one liked it. "*C'est dommage,*" said Dubrule, clicking her tongue with disapproval, and Clément agreed that it obscured the shadows of the lotus leaves on the white base.

They played with centimeters, centering elements, off-centering them, and Lamejer said, "Plexiglass is superexpensive," and Dubrule replied, "We'll look at our budgetary choices," but part of their calculation was actually the value of making it more, not less, expensive; for the brands that bring the most money into the store, Galéries puts their PLVs at eye level in the center of the store. The more money you put into advertising, the more the store does for you.

Dubrule stared critically at the PLV. "The stores only give us thirty-two centimeters on the counters."

Basset (sighing): "And this is thirty-five . . . OK." He looked at it. "It's the first time we've done it with so few images, but I think it's very pure." He brightened. "For the question of the expense, we can perhaps simply double them in the window."

Lamejer (critically): "We want the same thing on the counter and the window?" She looked around the room. No, apparently they did not.

Basset moved the layers of plexiglass. Dubrule made a delicate face. "Ehh ... *un peu trop de symétrique.*" A bit too symmetrical. He moved it back, and they stared at it, searching for their next move.

When that meeting ended, at 11:30 A.M., Dubrule and Clément immediately began another with Ellena, who had come up from Grasse. Clément began with a bright, if exhausted, smile. "The *concentrée* is going to be delivered today!" Dubrule let out the breath she was holding. Ellena looked a bit pensive. Clément was searching through papers, saying, "It matures in one week according to what Jean-Claude recommends, we put it in alcohol, we ... um ... let it macerate fifteen days, so the juice will be ready"—she looks up from the papers—"the fourteenth of October." Then bottling and shipping, so the initial delivery to them would be around October 21 or 22.

They each made some rapid mental calculations. "The eighteenth we have the duty-free presentations," said Clément, "which start in Cannes, and Rosanna [Rubino, who trained all Hermès sellers around the world in selling each new perfume and thus lived on airplanes] is going to start the training for the sellers. The press launch is November."

"So," said Ellena, thinking it through out loud, "maceration the fifteenth of October and available the twenty-first or twenty-second of October." It was clear he considered this short, though workable.

Dubrule: "Which means press samples available early November."

They were having timing problems, were behind schedule. Neither Bourgeois nor Ellena was happy with the limited time for letting the stuff macerate, the materials soaking together, balancing

THE PERFECT SCENT

and integrating. They discussed this, the words slightly tense, the ends of a few sentences stepped on now. If they worked it right, there should be just enough time. Clément found a conversational gap and proposed that for the press launch, instead of giving journalists real bottles of *Un Jardin sur le Nil*, they could use *factices*, bottles filled with colored water. They'd send them the real thing later.

In a pinch, added Ellena, who *really* didn't like the idea.

Clémerr registered his look. *"On fait une toute petite macération,"* she said, trying to soften it. We'll just do a very small maceration. "Four liters. We're jammed on the timing." She moved to a question about the pump. "We're going to test several kinds for reliability. The change of concentration could give us a little problem with the eau de toilette. We're really at the limit." She gave various alcohol percentages, how they'd manage things. "We'll do what we did with *Eau des Merveilles."* She looked up at Ellena.

"It's the *genièvre* [juniper] incense I used," said Ellena briefly. "It doesn't dissolve well in alcohol." These are the things that cause problems, because it's beautiful stuff, but it can create hell in the manufacturing.

Dubrule was looking at him. "Seriously?" she said. "There's incense in the perfume?"

Ellena: "Not a lot, but yes."

Dubrule realized, eyes widened: "The incense! . . ."

The maker had just sent the sticks of *Nil* incense that would go into the *coffret* box set. She got them, and they lit an incense stick, let it burn, inhaling the rich scent, slightly thicker than its liquid counterpart, filling the conference room. They watched the opalescent smoke curl upward.

<hr />

By October 4, Ellena was installed in his new lab. He'd spent the summer scouting for it. Gautier had given him free rein and deep pockets. He chose a 1960s starkly modernist concrete James

Bond–style house, utopian Corbusier architecture, all glass walls and cool ninety-degree angles like a modern artwork you lived in, just down the hill from the medieval town of Cabris, about five kilometers from Grasse. The house sat in a clean pine forest. He walked from the car to the front door through the scent of the jasmine growing beside the steps, and on clear, beautiful fall days like this one could look from the living room—which was where he'd placed his desk, a smallish simple wood table, at an angle to the walls— down to the Mediterranean laying itself against the Côte d'Azur. The chairs were real ostrich skin. Outside, the gardeners maintained things; he was going to have them put in orange irises, the Hermès orange. He could slide open the immense glass walls and let the fresh air sweep through. It had taken time—he'd skipped his summer vacation—but by late September Gautier had flown down from Paris, beheld what he had created, and pronounced it very good.

In the lab, Ellena had installed around 220 ingredients. He'd placed the lab in one of the house's bedrooms, though in any case the modernist architecture made all the bedrooms, and the living and dining rooms, feel like labs. He had:

Kephalis: a synthetic that smells like brown butter plus shag carpeting.

Left carvone: a synthetic that smells of chewing gum.

Cyclamen Aldehyde: the smell of cold.

Manzanate: compote of slightly overripe Fuji apple.

Leaf Acetal: a drop-dead gorgeous smell: marine plus green peas.

A synthetic that smells richly of olive oil. (It's actually a flavor.)

A mixture of synthetics sold together as a truffle scent. (Truffle smells mostly of a molecule naturally created in the fungus called dimethyl sulfide.)

A base (a mixture of synthetics) called Lait Concentré, Concentrated Milk. Mix this with Aldehyde C-18, and you get daiquiri.

Sigaride: a synthetic that smells like Fire Island Pines, the ocean/pine/sand smell of island plus the warm, slightly sweaty richness of the smell of the men. A mesmerizing scent. (The problem with the molecule is that it has very little tenacity.)

Decalactone Gamma: a synthetic that smells of milk without cream.

Geraniol Super: a wonderfully interplanetary rosy geranium. (The only thing that smells on a geranium is the leaves; the flowers smell of nothing.)

The average perfumer, said Ellena, would have a collection of around 1,000 materials, "but I don't even use the 220. Maybe 120. The others are there because..." He shrugged. "*Peut-être.* Maybe. One day. Like a word in a dictionary. You have it at hand. I have a tiny dictionary. I use few words."

Why.

"Why..." He thought about it. "I want to master what I'm doing. Mastering means that for each word, every material in the formula I know why it's there, and I control the formula. And behind that there is also the desire to show that the perfume is not the result of some chance but a reflection of a reasoned process." He made a very definite series of stepping motions with his hand, squinting at a target. "It's intellectual work. When you start out, it's more about your passions. At the end, it's intellectual. *Un Jardin sur le Nil.* It is, it has to be said, both this green mango, this fruit that I pick, and, in the same instant, all the images that sets off. That, that's passion. And from that moment, I ask myself: How can I transform that into a perfume? Can I tell a story there? This is a scent spun into a story. And after, there's a *décodage,* a synthesis, and after that, I have fun. Putting this and that in. The melon. The grapefruit."

He paused. "Today, as far as I'm concerned, *Un Jardin sur le Nil* is finished. I'm just tweaking a few things, just for the pleasure of doing it." Just to see what happened. He thought about it. "At Hermès, I'd like in the future when you go to Pantin that there would be an olfactory curiosity cabinet. That guests I meet there would be able to smell not only what I've done for Hermès but things that will never be finished. Just for the pleasure of it."

He and Gautier were leaving in a few weeks for Aswan for the press launch at the Old Cataract Hotel.

⁓

They left Paris on Egyptian Airlines, a six-hour flight, on November 13. The start of winter. The last time Ellena had been in Aswan, it had been spring. Now much had transpired, and he was sitting with Gautier, Dubrule, and Olivier Monteil. This journey would pave the way for press trips to come, many launches of important perfumes. The juice was done, the bottle, the book, the packaging. The captions for the photographs were approved. After landing, they checked into the Old Cataract Hotel, and Ellena walked down the corridors again, soft and dry with age, and felt at home. Out the large windows, the Nile lay black and gleaming and glass-still, a black shiny cobra sunk into the desert sand. The ruins that were still lit every night for the viewing pleasure of guests.

On November 15, the first group of journalists arrived, European and North American press, escorted in by Francesca Leoni, the head of Hermès's North America communications. Four Italian journalists, a fashion writer from Madrid, Eva Chen from *Elle* in New York, and a few others. Everyone would sail up the Nile to a lunch on the island where they'd discovered the green mangoes. Monteil, the press handler, had arranged everything. He was executive producer of this show, and he had worked out the details, done all the blocking and produced the scenes.

The journalists came down the next morning, blinking in the light, for the first piece of the production. They wore dark sunglasses with designer names on them. Monteil and the Egyptian sailors guided them from the dock into a wooden boat—not a felucca but an ordinary wooden transport with a motor and a roof to shield them from the sun—and they set off. They passed the Nile's biblical reeds, moving gently in the cool black water. They gazed at the giant cakes of stone and towering dunes of sand frozen in on-rushing motion. They passed the Club Med. Boats full of Egyptian families cruised by, the kids shouting "Hello! Hello!" The journalists waved back desultorily.

The Italian writers wore immaculate white cotton shirts and slacks and held their Chanel bags and smoked and gazed at the backdrop of the book of Exodus. Eva Chen was fresh and pert, looking around with interest. Leoni was wearing a $1,200 pair of Hermès sandals and head-to-toe ethereal white linen. The Spanish fashion journalist, an obese woman with a mustache, wore huge, jet-black sunglasses with the word *GUCCI* written in gigantic letters on both sides. She smoked one cigarette after another. As she finished each one, without any indication of giving a thought to it she casually threw the spent, tar-stained butt on the surface of the pristine Nile, where it floated, a white/gray speck of trash, toward the reeds and the herons' nests, and lit another, inhaling the filth deep into her lungs. Her thick yellowish skin stank of charred cigarette.

The Egyptians pulled the boat up in a tiny cove on a tiny island. The journalists looked up, and Ellena was already waiting for them on the island, sitting in his chinos and white button-down, on a rock, like Rodin's Thinker. "We looked for an island with just one mango tree," he said, turning around to gaze at it for a moment, "and we found it!" (They'd scouted for days, Monteil scouring the islands in the river.) Hermès had told the journalists nothing about the perfume. Ellena gave them a few words about the tree, the fruit, the scent. It was not mango season now, and there was no fruit on the tree this time. Imagine, he said, and began to tell them. By the time he was done, they were thinking of Aswan in the spring, the hot sun and, on the lower branches, round globes of dark green fruit. There was a sound behind them, and they turned around. The transport had disappeared, and in its place magically was an orange, green, and white Egyptian felucca, its white sails luffing in the breeze.

They boarded. Ellena sat in the middle beside a cooler of drinks Monteil had carefully filled. The journalists selected cold waters, Cokes, lemonades. They glided downstream and listened to him.

Ellena: "Before coming here, I had already finished the perfume in my head. It was based on all my preconceptions of Egypt and the desert and the Nile, and when I arrived I realized I was completely wrong. I was imagining orange flowers and jasmine, heavy smells. I didn't find them in the air here at all. Wrong. So I spent two white nights, truly, *deux nuits blanches,* and it was difficult to throw away my idea, my preconceptions. Throwing them away was a lot of work."

He was interrupted by his cell phone. Everyone laughed. The Egyptian sailors peered in to see what the sound was. He struggled to turn it off, finally succeeded. Rolled his eyes and stuck it back in his pocket.

"I smelled cassie, hibiscus, lotus—and then mango." A desert falcon veered wonderfully from bank to bank, and everyone including Ellena watched its trajectory, holding their breath. The wind filled the sails, dry, warm, and lovely. Ellena took out the *touches.* Carrot, said Ellena, handing out the scent, and they smelled, wordless. Fruit, he said, passing around another *touche.* Green. The lotus he got from the souk (someone dug into the cooler, and he shifted for a moment), where the sellers put the root in water and, he said, get a smell halfway between peony and hyacinth.

He showed them the small orange notebook he used to take notes, opened the pages and held it up to them so they could see his original formula (though he made no move whatsoever to actually hand it to them; he closed it back up and carefully put it away). "I don't use head space," he said, the registering of complex natural scents that are then reproduced meticulously in a lab to re-create exactly the same mix of molecules. "That's fact. It's *literal*." He shrugged off the literalistic approach. "I use imagination, which is much faster than any machine. There are three to four hundred molecules in the smell created naturally by mango. I use four. I believe the illusion of mango is stronger than the reality. This is a matter of commercial seduction. I use your own memory to induce a reaction from you, triggering your desires and pleasures and thus associating them with the perfume. For the perfume things go very slowly. We spent four days in our felucca just catching smells. The difficulty is in choosing good from bad."

An Italian journalist asked him if the perfume, which they still had not smelled, was a unisex. "I don't like the word 'unisex,'" said Ellena. "That's no sex." She thought about it. He watched her, sought to explain his point of view. "For me, if you write a book, it's not for a man or a woman. It's a book." She nodded. "For anyone who loves to read. I can wear *Shalimar*. If I like it, I wear it."

And then it was time. He took out the finalized, fully macerated perfume and sprayed the *touches* and handed them out. The journalists took them in the way that fashion journalists do, as if it was their right, and with guarded looks. They smelled. They gave away nothing. They glanced at each other behind the sunglasses. They sent brief appreciative looks at Ellena, but really they were playing their cards close. No one said anything. No one asked to wear the scent on skin. The Spanish journalist alternated sniffs of her *touche* with deep drags on her cigarette. Her thunderheads of gray smoke drifted over the Nile's dark, clear water and curled around the *touche* she clutched, drowning Ellena's perfume.

"Classical perfumery," Ellena said to them as they held the *touches*, "is too perfumey for today's sensibility. It's like reading Balzac. This is a new way to write it." He told them how in *First* he'd put in four different jasmines, three roses, two lilacs. Redundance, he said, complex and obvious. "I like Ravel, Debussy," he said.

"Matisse." He couldn't do *First* anymore, he explained. In *Nil* he knew that he meant to put his paintbrush here, and here. Not here. He glanced at the glass bottle of *Nil*, the classic Hermès shape so precisely colored in greens and blues, and they followed his gaze.

Evening. Dinner in the Old Cataract's Presidential Suite, an elaborate meal. Hermès had flown in a Frenchwoman named Véronique, a stout, pleasant woman, just to create the table decoration. Everyone was received on the terrace overlooking the pool below, the giant palms bathed in electric illumination below that, and the Nile below that. Véronique was worried that the Saharan wind would blow out the candles, which it did. As journalists entered past the hotel's Egyptian waiters, they saw for the first time the bottles and the *coffret prestige*, the book of Ellena's text and Bertoux's photos, and the packaging Dubrule had created. There were four leaves on the cover. The 100 ml and 50 ml bottles. Gautier and Ellena engaged in a rehearsed and wincingly awkward exercise in which Gautier asked questions and Ellena and Bertoux responded, all in English, although Gautier's substitution of French words in English—"So, during this *voyage*, this—*promenade* in *Aswan* ..."—found the Italians staring off over the Nile or ordering coffee at length from the waiters. (Gautier to the journalists: "What is your favorite photo?" The journalists glanced around disinterestedly at the enlarged copies of Bertoux's photography, which Hermès had so carefully created and installed all around the table, then picked at their desserts.)

While everyone was eating, Monteil got the keys from the front desk and delivered to each journalist's room, placing it with care on the bed so it would be the first thing their eye caught on coming up from dinner, the *coffret*: a 100 ml flacon of the fully macerated A* (in the Saint Gobain bottle with the blue-green-yellow glass *vernis* and the silk-printed *Hermès* exactly nineteen millimeters from the bottom), the book (*Rencontre, au fil du Nil*, text Jean-Claude Ellena, images Quentin Bertoux), and the sticks of Japanese incense (fully intact). On a small green paper lotus leaf were written, in gold ink, the words *Un Jardin sur le Nil.* The bottles of perfume were laid on the laundered, pressed, cool thick cotton sheets of the beds.

⤳◎

The Hermès team did more press, and still more. Three waves of journalists, different media markets, a group flown in from Asia, and Ellena had already said everything, told the story so many times. Presented his creation.

They flew back to Paris after it was all over. Ellena wondered on the flight if he'd said everything right. He wondered if he'd put everything into the perfume that he should have. He considered it a perfect perfume. He considered it desperately flawed. He could have done it better with more time. He couldn't have done it better, not with all the time in the world. He smelled it in his head.

When he got home to Grasse, exhausted, Ellena slept, got up the next morning, and went to his lab in the modern house. In March, *Un Jardin sur le Nil* would be in stores. At his desk, he sat and looked out at the Côte d'Azur below him, then started to work. He had the beginnings of several possible *Hermèssences.* And he was working on some other ideas. One was the smell of a leather bathing suit emerging from a swimming pool. And there was a leather sprinkled with sugar. He wasn't worried. For the moment he was floating, smelling, looking for something. Maybe the answer was already there.

C OTY LAUNCHED *LOVELY Sarah Jessica Parker Liquid Satin* in
September 2006. It entered the market and joined
Lovely, which was consistently one of the best-selling
celebrity perfumes on the List.

But Walsh, Timiraos, and Parker were already heavily involved
in developing Parker's new perfume. Even as they were launching
Liquid Satin, they had created the new brief, and they went out with
it to the various Big Boys. The perfumers parsed the brief's concepts, pored over the clues it offered, the words and ideas Parker
had used to express this next new scent she envisioned. Then they
opened their palettes of molecules and began putting together their
first *essais*, their olfactory sketches. These they sent back to the Coty
team, which huddled over the scents.

This time IFF did not win the brief. Coty selected the submission of Frank Völkl, a perfumer with IFF's competitor Firmenich.

Coty had, for this perfume, engaged the well-known creative consultant Ann Gottlieb, an icon in the industry who had helped create *cK One*, and Völkl worked with Parker, Gottlieb, and the Coty team to create the juice. Chad Lavigne again did the bottle design, and, again, Trey Laird did the images.

Covet, Sarah Jessica Parker's second perfume, launched in August 2007.

〜◎

Hermès launched *Un Jardin sur le Nil* in March 2005.

With *Nil* finished, Ellena turned his attention to the *Hermèssences* collection, which had debuted with four scents in 2004. In November 2005, Hermès launched Ellena's fifth *Hermèssence* scent, *Osmanthe Yunnan,* which was his concept of an osmanthus flower he smelled in Beijing, and in September 2006, the sixth, *Paprika Brasil.* He also created what he considered as a perfumer one of his most important masculines, *Terre d'Hermès,* which the house launched in 2006.

In 2007 he created a perfume of a singular importance to Hermès. It was called *Kelly Calèche.* Hermès launched it in September. Though it used the *Calèche* name, it was not olfactorily a flanker; Ellena had created an entirely new perfume. The genius of *Kelly Calèche* is that it opens on the skin as a transparent modern, built on a sunlit green that approaches but, with a delicate exactitude, does not actually touch floral. The fragrance is structured by a glass angularity whose beauty is as much in the precision of its calibration as in its scent, a highly architectural piece. But it is only after a certain time that one realizes where it is going. The perfume becomes that rarest of things, a completely wearable contemporary leather.

O NE OF the greatest pleasures of writing a book is that you get to thank people. I have a lot of people to thank.

This book originated in two articles, the first of which I did for *The New Yorker* on the creation of an Hermès perfume. My thanks to my editor, Daniel Zalewski, and to David Remnick, who proposed this idea to me—and to Louisa Thomas, for her ineffable fact-checking, to Elizabeth Pierson-Griffith for her sang froid in the face of some eleventh-hour trans-Atlantic panic, and to Stanley Ledbetter for his lovely good humor. And the second, on Sarah Jessica Parker's perfume, for *T: The New York Times Style Magazine*, which my editor, Andy Port, expertly helmed. Andy edits my "Scent Notes" column, offers crucial help and support, and is a constant model of grace under fire.

I reported much of this book during my first two years writing for *The New York Times* on perfume and the scent industry and then as its perfume critic. My thanks for this ongoing class go to Alix Browne, Jamie Wallis (who is missed), John Hyland and George Gustines, Diane McNulty, Pilar Viladas, Kara Jesella, Horacio Silva, Gerry Marzorati, Trip Gabriel and Mary Ann Giordano, and Justine Simons, Ann Derry, and Kassie Bracken (a sorceress with an editing machine). *Times* fact-checkers have saved me repeatedly and smoothed out many last awkward clauses: my thanks to Adam Kepler, Andrew Gillings, Lia Miller, Andy Gensler, Ian Keldoulis, Renee Michael, Ursula Liang, Sahara Briscoe, and John Cochran.

I am deeply grateful to *The Times*'s Jim Schachter, reliable guide and counselor extraordinaire. Pat Eisemann has gone above and beyond. Mike Oricchio is svelte and Lad Paul is gorgeous (or vice versa, who can remember?) and to both I owe a bundle. You can't do better in Paris than hanging out with Elaine Sciolino. And Janet Maslin seems to be a focal point of innumerable marvelous things.

Above all at *The Times* I'd like to thank one person, Stefano Tonchi, the Style Editor of *The Times Magazine*. It is Stefano to whom I proposed the idea of a *Times* perfume critic, who got it instantly and believed in it completely, who made it happen, and who's been unfailingly supportive. Thank you, Stef.

Bretly Thorn, diner extraordinaire, did some extremely useful on-the-spot test reading. Francesca Leoni introduced me to Stefano at a party. And getting to know Tom Bezucha in the course of this project has been an unexpected pure pleasure.

Michael Edwards's professional perfume database www .fragrancesoftheworld.info is as crucial a tool for my reporting as it was for this book, and Michael's historical perspectives are represented on these pages. Giovanni and Anne-Marie Luppi I thank for their expert *servizio linguistico*. My appreciation also goes to Diane Nicholson, Leslie Singer, and Karen Grant at NPD.

Those in the perfume industry—perfumers, creatives, evaluators, execs, all of you too numerous to list here—who helped me in ways large and small—have my deepest appreciation. You have my sincerest admiration as well. I want to mention Céline Ellena's contribution, both for her familial memories and, more, for her expert, precise rendering of her olfactory work into verbal language. And Givaudan's Dr. Philip Kraft, whose knowledge of perfume chemistry is breathtaking and who was instrumental in helping me avoid some scientific errors; those that remain remain despite his efforts.

I would know nothing of this industry, and indeed most fundamentally I would never have had the idea of being a critic of its products without Luca Turin, who via a chance encounter guided me into this strange world.

This book was unusual in the degree to which it relied on the faith of its subjects. It takes courage, in the most real sense, to participate in a project of this kind. Aside from a few very basic ground rules, Hermès, Coty, and IFF did their due diligence, discussed the pros and cons internally, and decided to open their doors to me. They retained no controls, nor did they ever request any. If on a purely journalistic level I owe them nothing—they went into it with their eyes open—in the fundamental sense of their simple agreement I owe them everything. This includes everyone at those three companies, but I would like to mention by name Catherine Walsh, Carlos Timiraos, Belinda Arnold, and Sarah Jessica Parker; Clément Gavarry, Laurent Le Guernec, Melissa Sachs, Yvette Ross, Steve Semoff, and Nicolas Mirzayantz; and Bob Chavez, Olivier Monteil, Stéphane Wargnier, Anne-Lise Clément, Quentin Bertoux, Véronique Gautier, Hélène Dubrule, and Jean-Claude Ellena.

I wrote this book during one intensive October in two seriously terrific places in Rome: the apartment of my good friend Laura Tonatto and one of the best restaurants in Rome, which happens to be across the street from Laura, the marvelous Il Corallo on the Via del Corallo near the Chiesa Nuova. (It was there that Marco taught me how to talk Romano: "Chandler! *Di ho suonado brimma!*"). I sat at Il Corallo's tables on the ancient stone street under the umbrellas, plugged my computer into the outside electrical outlet, and worked under a cobalt Roman sky. When you go, say hi to one of the three owners, Anna, Orietta, and Antonella, and tell them thank you again for me. Start with the insalata Corallo and a basic *pasta cacio e pepe* and go from there. The food so good it will make you cry. *Grazie a tutti*—Sonia, Salvatore, Mauro, Carlos, Jimmy, Fedi, Dalila, Masud, Carlos, Giampiero, Carlo, Gabriele, Andrea, Anwar, Roberta, Arianna, Alonso, Ale, Alessio. *E Mauro, ricordati che sono io che faccio il migliore latte macchiato di tutta Roma!*

In New York, my support team was Michael Strong and Yorick Petri, with Joe Tomkiewicz and Richard "Toastie" Trost batting cleanup from Houston and San Francisco: givers of phone calls, senders of e-mails, takers of meals, offerers of criticism and medical advice and interventions, total pains in the ass, and of course sources of love. And to Razib Kahn, who is my Web site creator, maintainer, and all around genius and indispensable help, you rock.

Editing a book is an exercise in humility, and sometimes in abject debasement. Vicki Haire, my copy editor, did everything she could, precisely and with marvelous clarity, to help me avoid looking like a complete idiot. Eadie Klemm was as always just *uberschön.* That's got to be a word, Eadie. John Sterling and Jennifer Barth crucially fired the starting gun. Eric Simonoff, my agent, steered an immaculate ship. Sarah Knight has thrown herself into this with heartwarming dedication. George Hodgman, my editor at Henry

Holt, has been insightful and demanding and infuriating and supportive and implacable and should, for the job he did on this book, be sainted, as George himself would be the very first to point out. I don't know if his dedication or his expertise is greater. It doesn't matter; both are awesome.

ACKNOWLEDGMENTS

It is the first time I meet her. We're back at her house in the West Village, and we've spent the August afternoon together, and it has been blistering hot with a blue sky, then it rained violently under gray-black thunderstorm clouds, so now the smell of New York's streets is extracted and steamed into the air. Jennifer is packing the last of her camera equipment, and SJP is seeing me to the door. She gives me a hug. She hesitates.

"You know," she says, "creating the perfume?"

Yeah?

She's trying to place the impulse for this thing she's made in context. She wants to explain something. She says, a bit tentatively, "It was about that instant just after you spray it. This silvery mist in the air, and you're watching it settle on you? And you're holding your breath, really hoping, *hoping.*" It is an act of optimism, or of faith. Her face is focused, narrowed on the materials constructed from the formula she has in her head that have settled on the arm of her imagination. "And you bring your wrist up to your face like this, and you close your eyes and—*there.*" She opens her eyes, surprised blue. "That's it."

After a second she focuses on me again, says, "It was entirely about that moment."

ABOUT THE AUTHOR

CHANDLER BURR is the scent critic for *T: The New York Times Style Magazine* and the author of *The Emperor of Scent: A Story of Perfume, Obsession, and the Last Mystery of the Senses.* His first book was *A Separate Creation,* about the hunt for the biology of sexual orientation. Burr, who earned a master's in international economics and Japan studies from the Paul H. Nitze School/Johns Hopkins, has written for *The Atlantic, The New York Times Magazine, U.S. News & World Report,* where he was a contributing editor, and *The New Yorker.* He lives in New York City.